T0323909

Practical Root Cause Failure Analysis

Root Cause Failure Analysis (RCFA) is a method used by maintenance and reliability industry professionals as one of the key tools to drive improvement. This book offers a quick guide to the applications involved in performing a successful RCFA by providing a foundational view of maintenance and reliability strategies. It also highlights the practical applications of RCFA and identifies how to achieve a successful RCFA, as well as discussing common equipment failures and how to solve them. Case studies on topics including pump system failure analysis and vibration analysis are included.

This book also:

- Suggests examples on how to solve common failures on many types of equipment, including fatigue, pumps, bearings and mechanical power transmission
- Highlights practical applications of RCFA
- Identifies key elements for how to achieve a successful RCFA
- Presents case studies on topics including pump system failure analysis and vibration analysis

This book is a must-read for any reliability engineer, particularly mechanical reliability professionals.

Reliability, Maintenance, and Safety Engineering: A Practical Field View on Getting Work Done Effectively

Series Editor:
Robert J. Latino, Reliability Center, Inc., VA

This series will focus on the "been there, done that" concept in order to provide readers with experiences and related trade-off decisions that those in the field have to make daily, between production processes and costs no matter what the policy or procedure states. The books in this new series will offer tips and tricks from the field to help others navigate their work in the areas of Reliability, Maintenance and Safety. The concept of 'Work as Imagined' and 'Work as Done', as coined by Dr. Erik Hollnagel (an author of ours), this series will bridge the gap between the two perspectives to focus on books written by authors who work on the frontlines and provide trade-off decisions that those in the field have to make daily, between production pressures and costs...no matter what the policy or procedure states. The topics covered will include Root Cause Analysis, Reliability, Maintenance, Safety, Digital Transformation, Asset Management, Asset Performance Management, Predictive Analytics, Artificial Intelligence, the Industrial Internet of Things, and Machine Learning.

Lubrication Degradation
Getting into the Root Causes
Sanya Mathura and Robert J. Latino

Practical Root Cause Failure Analysis
Key Elements, Case Studies, and Common Equipment Failures
Randy Riddell

For more information on this series, please visit: https://www.routledge.com/Reliability-Maintenance-and-Safety-Engineering-A-Practical-Field-View-on-Getting-Work-Done-Effectively/book-series/CRCRMSEGWDE

Practical Root Cause Failure Analysis
Key Elements, Case Studies, and Common Equipment Failures

Randy Riddell

CRC Press
Taylor & Francis Group
Boca Raton London New York

CRC Press is an imprint of the
Taylor & Francis Group, an **informa** business

CRC Press
Boca Raton and London
First edition published 2022
by CRC Press
6000 Broken Sound Parkway NW, Suite 300, Boca Raton, FL 33487-2742

and by CRC Press
4 Park Square, Milton Park, Abingdon, Oxon, OX14 4RN

CRC Press is an imprint of Taylor & Francis Group, LLC

© 2022 Randy Riddell

ISBN: 978-1-032-16465-6 (hbk)
ISBN: 978-1-032-16466-3 (pbk)
ISBN: 978-1-003-24867-5 (ebk)

DOI: 10.1201/9781003248675

Typeset in Times LT Std
by codeMantra

Contents

Foreword

There are many books written about Root Cause Failure Analysis (RCFA), so what makes this one different? As its title conveys (as well as the book series), the difference is PRACTICALITY! This is not a book written by a researcher or an academic, it is written by a seasoned practitioner, Randy Riddell. This means it is based on field experience, where an operational plant in a real-life working environment forms Randy's "lab."

For those of us fellow RCFA practitioners, it is refreshing to read a perspective that doesn't sugarcoat these realities but faces them head on and tells us how they have been overcome. We all are acutely aware that we do not work in perfect settings and not everything we simulate in a lab can be reproduced in real life.

For instance, a bearing may run forever in a lab setting, where operating conditions are always ideal for the design of that bearing, and the perfect amount of the correct lubricant is maintained.

However, we know the reality is those bearings are going to be exposed to imperfect operating conditions, as well as to imperfect people and systems that will purchase, store, commission, operate and maintain them. This exposure to true reality will dramatically impact the bearing life. These are all variables which didn't exist in the lab. How long that bearing will really last is up to us!

Practical RCFA is about using evidence-based, analytical approaches to uncover these realities and implement effective corrective actions that will strengthen the supporting organizational systems to prevent recurrence. But the goal of RCFA itself should not just be to prevent recurrence, but to organically create and share knowledge across an organization. When this does NOT happen, think about the amount of RCFA re-work (costs) that occur simply because we did not know that a problem we have has already been solved.

To this end, Randy starts off with laying the firm groundwork for what a *holistic* RCFA approach looks like. This focuses on the broader systems view, like:

1. The four levels of failure and the appropriate depth of analyses
2. Proactive versus reactive analyses
3. Exploration of physical, human, and latent root causes
4. Where RCFA fits in an effective Reliability system

5. The 5 'Rights' of a successful RCFA system
6. Using RCFA to create a knowledge management system (what I call "institutionalizing corporate memory")

It's one thing to understand principles, but it's another to successfully apply them. From this point on, Randy practices what he preaches, and demonstrates the application of these principles (the Physics of Failure) across many types of failure, such as key factors in:

1. Fatigue Failure
2. Bearing Failure
3. Pump System Failure
4. Mechanical Powertrain System Failure and
5. Bolted Joint Failure

It becomes evident the author is not only an effective RCFA facilitator in the field, but also quite an adept Subject Matter Expert (SME) in the technical aspect of the failures he uses for his extensive case studies. This is a unique and valuable combination.

Those who are practitioners will find this to be an invaluable reference as they develop their RCFA facilitation skills. Those who support RCFA efforts will truly understand the effort it takes to conduct a proper RCFA, thus giving them a new appreciation for how best to support the analysts. Systems integration professionals will gain a unique perspective of how they can integrate the successful logic from an RCFA into their knowledge management systems, making it easier to share lessons learned across the organization (and grow RCFA knowledge databases).

There is much to be gained from properly digesting this information, conveying it to knowledge and eventually converting it to wisdom. But in the end, it is up to us as to how far along the spectrum we will go.

I think this saying is appropriate here, as I view this as a paradigm that all of us in the RCFA business are united in our efforts to defeat:

We NEVER seem to have the time and budget to do things right, but we ALWAYS seem to have the time and budget to do them again!

Thanks for this significant contribution to the RCFA field, Randy!

Robert J. Latino – Principal, Prelical Solutions, LLC.

Preface

The reliability field is one of the most exciting fields to work and can bring great satisfaction in solving some of the most difficult and chronic failures. If you work in maintenance, reliability or engineering for almost any industry, root cause failure analysis is a key activity to improve your operation. Modern day industry has molded our manufacturing organizations with roles and responsibilities which embrace and drive RCFA activities. As many have stated before, reliability and especially RCFA are the most powerful tools for uncovering the hidden plant within our current operations.

Over the decades, organizations and companies have committed lots of resources to install RCFA systems and to transform their culture into a root cause finding culture. We have purchased training programs around RCFA, built massive databases, collected failure data, enhanced our computerized maintenance management system for failure analysis, added positions in our organizations to lead and manage RCFA systems and processes, and the list goes on and on. All of these are great things and necessary things to move our efforts to high levels of performance; however, it takes even more than all that to produce excellence in RCFA in our plants. Despite all our efforts and the popularity of RCFA, the results have not been a universal success.

For any mature reliability organization, RCFA is a core function. Additionally, anyone who invests in RCFA wants a successful RCFA, so why doesn't everyone achieve it? What does a good RCFA look like? In simple terms, a good RCFA finds the root causes and makes the necessary corrections to prevent or insulate future failures from occurring. More importantly, what are the key elements to achieve a successful RCFA? Guessing or being lucky will work sometimes but there is no need to depend on luck or chance for being successful. There are key elements that will greatly increase the chance for successful RCFA. The more key elements that are employed then the higher the chance that the desired outcome will result. This book will look at what successful root cause failure analysis looks like.

RCFA is a process – but mastering a process will not guarantee a successful RCFA. In three decades of using many different methods for RCFA, I've not seen a silver bullet in any of them. However, there are some good elements in each of them that can be key in helping to solve failures. This book does not attempt to teach a particular method or process but will utilize some elements of several methods to combine an overall RCFA effort to achieve RCFA success.

While the advancement of the IIoT and other condition based monitoring technology has simplified some of the human interaction with our industrial equipment, the need for human-driven failure analysis is still in high demand. Technology can determine the equipment conditions, give some fault indications, and provide good RCFA information but it can't determine root cause. Executing successful RCFA is still a human function so we must be equipped to execute it successfully.

Author Biography

 A native of Kossuth, MS, Mr. Randy Riddell attended Mississippi State University where he received a Bachelor of Science in Mechanical Engineering. He has over 32 years of industrial experience with a career focused on maintenance and reliability in the paper industry. His formal certifications include Certified Maintenance and Reliability Professional from the Society of Maintenance and Reliability Professionals, Machinery Lubrication Technician from the International Council of Machinery Lubrication, Certified Lubrication Specialist from the Society of Tribology and Lubrication Engineers, and Level 1 Vibration and Pump System Assessment Professional (PSAP) from the Hydraulic Institute. He has published many articles around industrial equipment reliability topics. He has also been a guest speaker at several industry and university engineering settings.

Introduction to Successful Root Cause Failure Analysis (RCFA)

1

Root Cause Failure Analysis (RCFA) is a reliability process used to determine the causes of failure of a piece of equipment, component, system or even process. The goal of RCFA is to find these root causes so that some action can be taken to eliminate reoccurrence or more feasible to reduce the probability of reoccurrence. This goal is the focus of this book to evaluate what are the key elements of successful RCFA not only by looking at general concepts but specific examples on real-world industrial failures.

I suspect that learning from failure has been something that humanity has studied since the fall in the Garden. The methodologies have certainly changed over time but the search for making things better has always been out in front of humanity.

I am reminded of a story about a young man who was being mentored by a seasoned businessman. The young man asked what was the key to his success? The businessman said, "Two words, good decisions." The young man continued to probe and asked, "well, how do I do that?" He answered, "One word, experience." The young man thought, I can't stop here so he asked, "Well how do I get that?" The businessman answered again, "Two words – bad decisions." RCFA turns into success when we learn from not only our bad decisions but the bad decisions of everyone else as well.

It probably needs to be said at this point that RCFA has several obstacles that keep us from wanting to learn from failure. For many the focus is on troubleshooting and not RCFA (more on that later). For some, firefighting is fun, and firefighting is rewarded by many organizations more than any proactive tasks. Probably the biggest hurdle in learning from failure is to refrain (avoid) from blaming others. RCFA should not involve blaming individuals or groups. Solutions will involve these two areas. While error is part of the human factor,

DOI: 10.1201/9781003248675-1

1

it is common among all of us. It has been said, "to err is human and to blame someone else, well that shows management potential."

I'm not certain as to when RCFA first started being used as a focused activity but I would guess that it became more mainstream following on the heels of the industrial revolution, like so many other engineering and technological advances. One thing is certain, it began out of necessity and after failure.

One such early failure occurred with the steam-powered riverboat, the Sultana. On April 27, 1865 on the Mississippi River near Memphis, TN three of four boilers exploded. The Sultana burned to the water line in 15 minutes. Over 1500 people died in the accident. The Civil War had just ended, and the nation was at the start of reconstruction and healing, so any form of RCFA was not completed and the root cause was never determined. In 1867–1868, there were 441 recorded boiler explosions. In 1880, there were 159 boiler explosions. Out of the need to prevent these repeating failures, a group of engineers met in 1880 and founded the American Society of Mechanical Engineers (ASME). ASME codes and standards began being developed for pressure vessels.[1] An obvious lesson here is that unsolved failures will become chronic. Once failures become chronic then they often become more impactful to the business or society and a higher priority to solve.

One of the early lessons of the damaging effects of resonance came after a very well-known catastrophic structural failure of the Tacoma Narrows Bridge on November 7, 1940. The structural stiffness of the bridge was such that as the wind blew across it sent the bridge into severe resonance until the bridge collapsed. Much was learned about resonance from that failure. Resonance is not a source of vibration but a magnifier of it. Resonance is also very destructive and must be avoided in our machines and structures to avoid failure.

1.1 TROUBLESHOOTING VS FAILURE ANALYSIS

Is there a difference between troubleshooting and RCFA? We see troubleshooting tables or charts all the time from equipment suppliers. We also see Troubleshoot, Cause, and Correct charts (TCC) but is that really RCFA? The American Heritage Dictionary says a troubleshooter is a worker whose job is to locate and eliminate sources of trouble. Well, that doesn't help much. While it has been debated, and I'm not sure there is an exact correct answer, here are some thoughts to consider. I believe there are two differences between troubleshooting and RCFA.

1. The end goal of the activity: Troubleshooting has an end goal of correcting the system or process so the line can resume production. The end goal of RCFA is to find the root causes of component failure and then to complete some action items to either eliminate or prevent future failure potential. Troubleshooting focuses on the cause of unscheduled machine downtime while RCFA focuses on improving the probability of future failure events of the same failure mode. Once the machine is back up and running the troubleshooting process is over but the RCFA process on the failed component begins. With the fact that the machine is down, troubleshooting becomes a necessary activity in which the organization does not debate or put off; it must get it done immediately. However, RCFA does not have the urgency by default as does troubleshooting, so only mature organizations will be successful in RCFA activities.

2. Analysis taken to the component level: RCFA eventually focuses in on failure at a component level while troubleshooting will only go deep enough to correct the machine malfunction. Troubleshooting is more of a reactive process, while RCFA is a proactive process for reliability improvement.

Here is an example of the two reliability tasks applied to a plant situation. The press section of a paper machine has a roll that will not load. The machine is down due to the issue so there is an urgency to commit resources to correct and get the machine back up. The goal is to get the machine back on production. Troubleshooting the problem is the focus of the team. Troubleshooting leads them to several directional valves in the hydraulic system with no solenoid power. A blown fuse is found and replaced. After changing the fuse and powering up, it blows again. Now the directional valve is the focus of the troubleshooting and is found to be locked up, which is causing the overload. The directional valve is changed out and the machine is started back up and in full production again. The troubleshooting task is now completed.

However, the RCFA task has not begun yet. In a purely reactive organization, it will not begin. It is off to the next troubleshooting activity on an urgent failure. The RCFA begins with having some component analysis of the locked up hydraulic valve. Inspection of the valve shows abrasive wear. Looking at condition monitoring programs, the last oil analysis for the hydraulic system was over 6 months ago (having skipped three samples) and the trend was showing that the ISO particle count was well above target on the hydraulic system. Further investigation revealed that the oil filter had a substitute part about 3 years earlier to save money and the filter, instead of being a 10µ, was a 25µ. Action items were put in place to systemize the oil analysis program and to change the filter element back to the correct filter. Additional training was also given for the technicians who operated and maintained the hydraulic system. Additional approvals were added to the process for changing store stock items to prevent technical mistakes with critical spare parts in the future.

Another example would be a suction roll vibration problem that suddenly occurred on a machine during one holiday weekend. The call came that the entire press section was vibrating severely. The vibration was so intense that it was vibrating loose many of the large (1.5") machine screws holding key components in the machine. The first part of the troubleshooting process involved determining which part of the machine was causing the vibration. Our vibration analyst was called in to diagnose the vibration on the machine press section. It was determined that the source of the vibration was the suction roll. After slow turning the roll and everyone looking the roll over, we could not find an external defect. However, the vibration showed a high 1X vibration and we suspected there could be a problem with the shell. The decision was made to change the roll. After the lengthy roll-change process, the machine was started back up and was running smooth again. This troubleshooting process was involved and very detailed.

The RCFA on the roll would commence over the next several weeks and the shell was found to be cracked in the middle about halfway around. The shell failed due to fatigue. While in some failures, troubleshooting may be the first part of the process, failure analysis is a separate activity.

These examples show the differences in troubleshooting and RCFA. Therefore, it is important to have clarity on the goal of the problem-solving efforts. Both are solving problems but have different goals for the outcome. Both are necessary to any manufacturing organization. Troubleshooting is the floor and successful RCFA is the ceiling when it comes to uncovering the hidden plant in every site.

1.2 DEFINING FAILURE – 4 LEVELS OF FAILURE

Failure could be thought of as a pyramid of categories where there are four levels of failure with Level 4 being the top and Level 1 the bottom as shown in Figure 1.1.

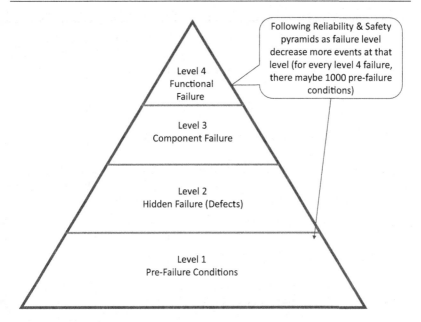

FIGURE 1.1 – Failure level pyramid

Level 4 – Functional Failure

What is failure or functional failure? One definition is when the asset is unable to fulfill a function to a standard of performance which is acceptable to the user.[2] This is not an emotional approval of the user to decide failure, but it is a matter of the asset design meeting design performance to a clear established standard. If a pump is to deliver 1000gpm (3785lpm) of fluid to the process as designed, then a functional failure would be when it can't deliver 1000gpm to the process. If an operations management decision resulted in wanting to get 1400gpm (5300lpm) from the pump system, then this would require a system redesign. It is a failure when changing the pump out would restore the function of 1000gpm.

If the pump was only able to deliver 700gpm, then it would be a functional failure. The pump may have wear, improper clearances, or some sort of line blockage. A failure in the pump that caused the functional failure would be a component failure. A component failure would be the lowest level of failure, which caused the asset to fail. For this example, it may be the impeller or suction wear plate. If the bearings or mechanical seal failed, these would be a component failure. Successful RCFA must focus the analysis on the component failure mode.

In addition to the functional failure example above, the pump could have a locked up bearing or sheared coupling where pump doesn't turn at all. This may be a breakdown failure where the asset has a broken part. All of these would be a functional failure, but the nature of the failures is more severe when a pump locks up versus wear where the pump has slowly lost performance.

Level 3 – Minor Failures

Minor failures would be failures where the loss of function of that component has occurred, but the function of the equipment or parent asset is still meeting performance standards. In the case of the pump, a minor failure may be that the seal is now leaking product. Here is a functional failure of the seal (component functional failure) but the pump is still producing 1000 gpm to meet its function at the asset level.

Level 2 – Hidden Failures

A hidden failure might be one that is a defect in the part, but it is still functioning, at least for the moment. The failure is hidden as far as the function of the asset is concerned but is a real defect on the part. An example here might be a bearing defect. The function of a bearing is to support the load while allowing the rotation of the shaft. That still happens with a bearing defect but eventually it will turn into a functional failure if allowed to go on for long enough. A lubrication breakdown that is not known is another hidden failure. A part with a fatigue crack growing is a hidden failure. These things may or may not be known or could have the ability to be known by some technology. This would be the area of the failure curve where failure is occurring, but it has not been detected yet.

Level 1 – Pre-failure Conditions

A pre-failure condition may be any number of issues that will contribute to or cause failure if left unchecked. An example of these could be misalignment, imbalance, cavitation, leaks, mechanical looseness, poor lubrication, corrosion, a bent shaft, resonance, high vibration, high temperature operation, and any other condition that is not ideal for the best operation of the asset. Some other items might be operating outside the operating design parameters for the equipment, such as too much pressure, too much flow, too much load, too much speed or too little speed. All these types of things can be preconditions or symptoms which can be contributing causes or root causes of failure. A pre-failure condition is not an actual failure. There are always changing operating conditions of equipment and systems. The equipment must run through these conditions, most of which can be corrected before a functional failure occurs or even a Level 2 or 3 failure.

All these failure levels lead to functional failure if not addressed. RCFA can also be triggered at any level of failure. If you can solve why the pump has high vibration (Level 1 pre-failure condition), then you will likely solve some of the root causes of the pump failure. RCFA is typically thought to be a reactive-type reliability activity and for much of the time it is. However, RCFA completed on Level 1 pre-failure conditions can transform RCFA into a proactive activity. Chronic failure RCFA could also be a proactive activity as chronic failures reoccur unless corrective measures are taken on the root causes.

The level of failure that is identified on the equipment can help in understanding the urgency of a response to that failure, as well as the things to consider when a failure analysis is executed for the eventual failure. The more proactive an organization may be, the lower down the failure level will their reaction to maintenance be. Consequently, the more reactive an organization is, the higher the level of failure is typically reacted to, which would be Level 4 or run to failure. Show me a plant with more Level 4 functional failures and I'll show you a reactive maintenance organization much of the time. The type of maintenance strategy may correspond as listed below.

- Level 4 Failure – Run to Failure Maintenance
- Level 3 Failure – Preventive Maintenance
- Level 2 Failure – Preventive & Predictive Maintenance
- Level 1 Pre-Failure Condition – Predictive Maintenance & Proactive Maintenance

Predictive maintenance is both reactive and proactive depending on the issue. Predictive maintenance that finds a bearing defect is reactive. The bearing defect has already occurred. Failure is not prevented, it is only managed when found early by vibration analysis. Vibration analysis did not prevent failure in this case, it only provided more time to plan for maintenance. Vibration analysis that identifies a Level 1 pre-failure condition, such as misalignment, would be proactive in the sense that if the pre-failure condition is corrected, a higher-level failure may be avoided.

Reactive predictive maintenance conditions would include all types of bearing defects, a bent shaft, and oil analysis wear metals. Proactive predictive maintenance conditions would include misalignment, imbalance, cavitation, looseness, resonance, and key fluid properties from oil analysis such as viscosity, additives, contamination, oxidation, etc.

1.3 RELIABILITY & RCFA

What is reliability? One definition is the probability that a product, system or service will perform its intended function adequately for a specified period or will operate in a defined environment without failure (~functional failure). For most rotating equipment, and even some static equipment, there is no such thing as 100% reliable. Industrial machines for the most part have a finite life. You could say they are designed to fail at some acceptable point in their service life.

Managers at NASA in the 1980s estimated that the space shuttle would have a reliability of 99.999% which is a failure rate of 1 every 100,000 flights. Engineers were a little more realistic and predicted a reliability of 99% (1 failure every 100 flights). That doesn't sound so good if you are an astronaut. The first space shuttle accident with the Challenger occurred at the 25th flight which was 96% reliability. In 2003, the Columbia incident occurred at the 113th flight. At the end of the shuttle missions, overall reliability was 98.51%.

The engineers were not too far off. Despite the many safeguards that NASA had taken the final product still had failures. Why is that? The human element is the powerful common denominator. Every failure has a human element to consider at every step along the way of the life of the asset. From managers' decisions to human resource management to purchasing to engineering to maintenance to operations. All human groups and their decisions affect the reliability and thus the failure of equipment.

To understand solving failures, it is helpful to understand a little about the life cycle of an asset. In the life cycle of an asset, it goes from design, installation and operation until its performance or condition deteriorates and maintenance is executed. This life cycle is often shown in the P-F curve or modified DIPF curve as shown in Figure 1.2. Point P is the point where a failure begins to occur or the point where its condition begins to deteriorate. F is functional failure or Level 4 failure. Other stages of failure may happen further back up the P-F curve. Every different type of failure mode has a different slope of the P-F curve region. Some may be steep, and some may be flatter. For example, a bearing that suddenly runs out of oil will have a steep and short P-F curve. A bearing outer race defect due to static corrosion may progress from a Level 2 to a Level 4 failure over many months so it would have a flat curve compared to a lack of lube failure mode.

On the front end, the asset has been designed and installed to do a certain function and reliability is also designed in. Design and installation determine an asset's foundational ability to perform at a certain reliability level. The length of the flat part of the DIPF curve will depend on how well the design and installation portions have been executed. Level 1 pre-failure conditions shorten the I-P

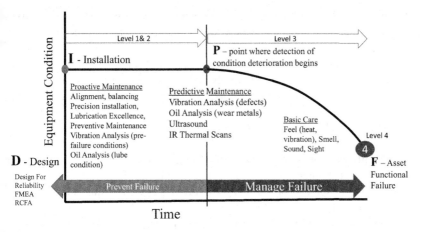

FIGURE 1.2   – DIPF curve

region and begin the P-F failure curve. Other factors after startup also affect how long the asset will run reliably. Once the equipment has entered a particular failure mode (P), the focus shifts toward managing the failure as it progresses. The main areas to prevent failure must happen prior to P in the design, installation, operation and maintenance portions of the equipment's life.

The result of that RCFA will address the different issues from the root cause of failure which may be something in design, installation, operation, maintenance, etc. The new or rebuilt asset is installed and the cycle repeats.

1.4 PREDICTING FAILURE

Predictive maintenance utilizes many tools to help maintenance identify and manage early-stage failures before Level 4 functional failure occurs. Whether the condition is a vibration, oil condition, ultrasound or temperature, there are some levels that are proven to lead to functional failure if left unaddressed. However, the area manager once he hears about a failure condition or precondition on his equipment, wants to know how much remaining useful life (RUL) he has so that corrective plans can be put in place.

How we respond will depend on several factors, such as the level failure (1, 2, 3, or 4), equipment criticality, the P-F interval for a particular failure mode, and the cost of unreliability (functional failure cost). Point P in the P-F curve is the point at which a failure condition can be identified. By the shape of the curve the condition continues to deteriorate at some slope until functional failure. At this point, assume that the condition monitoring process is 100% accurate. The next question is how long until functional failure? To find the answer, we must dig a little deeper.

Every failure mode has a different shaped P-F curve. For example, an inner race defect from corrosion will have a different P-F curve than that of fatigue spalling. Also, any other variable that is different in the same failure mode will also change the P-F curve shape. For example, the same inner race defect from corrosion, with a bearing that is lightly loaded compared to one heavily loaded, will have a different P-F interval. Consider the variables that can be different for each type of failure mode for a bearing and there could be hundreds of P-F intervals that would affect the RUL.

So, to put some good estimates on the RUL in predicting failure of multiple data streams that report on the condition requires a very good understanding of not only the equipment but the analysis of multiple data streams reporting on the condition – vibration, oil analysis, ultrasound, temperature, etc. Equipment items will be discussed in other chapters so a few thoughts here on condition monitoring tools.

Vibration has been called a science and an art. It is a science in the fact that all the programs, software and algorithms are based on the science of motion. These frequencies, generated by the rotating equipment, are based on the physics of the angular speed and geometry of the equipment design. When any of these thought-to-be-known parameters change due to wear or other physical changes, the science doesn't change but we don't have the new parameters to input into the analysis which results in flaws. For example, a piece of equipment has a bearing changed and the new bearing, while it fits, is a different brand with different geometry. Well, the defect frequencies are not going to be the same.

The other challenge with vibration analysis can be in reporting what the vibration shows versus what is causing the vibration. For example, a high 1X vibration can many times indicate imbalance. However, a bent shaft can also show high 1X as it gives the same view as imbalance mass in rotation. The mass is not centered around the axis of rotation but if the vibration report says you need to balance the rotor or clean the rotor, the root problem will not be addressed.

Vibration is a mixed bag of reactive and proactive. Reactive results from vibration are bearing defects. The failure has already begun, and nothing can prevent failure at that point. However, vibration analysis can also identify pre-failure conditions (imbalance, misalignment, vane pass) that when corrected may prevent premature failure.

Temperature monitoring can be tricky. Small changes in temperature may not mean very much, whereas large temperature changes obviously mean something more serious is occurring. Usually there is little reaction time when temperature changes reach certain levels. Understanding the source of heat is critical on each application. Is there steam used or only mechanical friction? Is there lubrication to control friction and heat? Is it oil or grease? Where is the temperature measured?

Oil analysis can provide several layers of condition monitoring. First it provides Level 1 failure conditions when wear elements begin to show up in analysis. If the wear has already started to occur, then the failure has begun, which is reactive. It doesn't mean if the condition is not quickly addressed that wear can't soon return to normal levels. A more proactive mode for oil analysis would be to closely monitor the lube condition, such as viscosity, contamination levels, additive levels, oxidation, Total Acid Number which can reveal degradation in the lubricant. If lube issues are not addressed it will lead to machine wear and ultimately functional failure.

To say that predicting failure is difficult if not impossible is an understatement; however, with an experienced professional some educated guestimates are possible. If enough study has been done on particular failure modes on the same equipment, then some statistical data may be obtained on P-F curve intervals and RUL when failure conditions are discovered.

1.5 MAINTENANCE & RCFA

An asset has several stages in its life cycle from design, procurement, installation, operations, maintenance, and eventual failure. The failure and changeout should have an RCFA. If the RCFA finds a design issue or has an action item for a redesign the cycle starts over again. Every asset and process in a manufacturing plant begins with design. Engineers, managers, and all the decision makers decide what will be installed, how it will be designed, and how much reliability will be built into every operating plant. There are so many parts of the asset life that are baked in up front with design choices. I have seen estimates as high as 50–75 percent of reliability issues have some connection back to design. Design choices will affect all levels of failure from Level 1 through to Level 4. Is the baseplate or foundation sufficient to minimize pre-failure conditions? Is the equipment selection such that components reach full life?

Procurement really should be following the engineering specifications and industry standards. When going out for bids it is important to have these specifications closely reviewed and adhered to so that these are all satisfied. There is always pressure to find a lower bidder. A good rule is never to include a supplier on the bid list that you do not feel can meet the specifications of the project. Don't use substitute materials outside of the specification. Be careful going to knockoff suppliers who offer a lower price but have cut corners. Price and value are two different things. Value is paid for typically over the long haul while price is garnered one-time upfront. Spare parts are another critical element on the front end of design and procurement.

Installation has many pieces which may include choosing the right contractors on initial install or even during maintenance later as the equipment is replaced by a contractor. Proper checkout and commissioning are critical in passing the baton to the operations group for successful asset function and reliability. Have the maintenance staff been involved in signing off from critical installation parameters that would lead to pre-failure conditions such as misalignment, pipe strain, imbalance? Training for operators and maintenance is an important part of the installation and commissioning of the equipment.

Operations will only be as successful if the design, procurement and installation phases are successful. Proper commissioning for the first time as well as having established standard operating procedures (SOPs) for future startup is important. Centerlines for operators so each shift operates to the same standards. Training for operators so new employees is brought up with the same knowledge standards as the initial startup crews. Do operators understand the design limits of their equipment and operate below them? Do they routinely swap spares on schedule? Do they make regular equipment basic care rounds looking for pre-failure conditions? The operators are typically the most available resources to the operating equipment and must take ownership of the equipment if full life is to be obtained.

Maintenance can have a similar impact on the asset as the original contractors who installed the equipment. Maintenance should be performed on the asset. Does maintenance have the correct procedures and specifications necessary to keep the equipment in this condition? Is the planning and scheduling function such that this information is available for them at the time of maintenance? If not, then this is an area where failure can be introduced into the equipment. For example, having the wrong gap, wrong spacing, wrong alignment, wrong assembly, wrong bolt torque, or substituting the wrong materials can all lead to equipment failure. Is lubrication being executed as and when it is required?

Proactive maintenance will do more Level 1-type failure corrective work. When the equipment is misaligned, maintenance will do an alignment. When the seal starts to leak, maintenance will fix the seal before the entire asset reaches functional failure due to a seal leak. Anytime maintenance changes parts, an RCFA may be initiated. Most of the time it will not be feasible, but the opportunity exists to ask the questions – did we get full life from the asset or component? If not, then why not? What caused it to fail before full life?

When the RCFA is executed, it may lead to a redesign of the system or the area where the root cause of failure was determined, whether it be design, procurement, installation, operations or maintenance. The total reliability of the asset is the sum of all the people who had some hand in the asset as defined by the equation. All the people's decisions in these groups and others over the life of the asset affect the total reliability of an asset today.

$$Total\ Reliability = \Sigma\ Engineering + Construction + Operations + Maintenance$$
$$+ Purchasing + Reliability + Safety + Environmental\ Management$$
$$+ Senior\ Management$$

A management decision made by any of these groups a decade earlier can affect reliability today. I have often heard this after we have discovered that an activity being done has at the very least contributed to an asset failure or short life. The answer many times is *the area manager told us to do it that way 10 years ago or we stopped doing this proactive work because our budgets were cut.*

With all the things maintenance must execute, how much should be devoted to improvement (RCFA)? Society of Maintenance and Reliability Professionals standards would show 10% for continuous improvement efforts, which includes RCFA. Eighty percent of the workload should be from repetitive work (preventive and corrective maintenance) or driven from preventive maintenance. That only leaves 10 percent for break-in work during a planned work week, which would include some unplanned failures.

However, if an organization is suffering from high Level 4 failures, run to failure maintenance strategy or reactive type maintenance then a higher focus must be on RCFA. RCFA then becomes the engine that moves the plant condition from reactive to proactive, from failure to reliability. The failed equipment must be replaced in order for the plant to operate, but the only way to get out of the rut is to determine the root cause of the failure and successfully execute the action items. If the cycle does not have some change in it, then it will repeat. When it repeats, chronic failure is the new norm. So, what must be done right for successful RCFA?

NOTES

1 Debbie Sniderman,The Greatest Maritime Disaster in US Hisotry, January 18 2011, ASME.org.
2 John Moubray, RCM II, 2nd ed, Butterworth Heinemann Publishing, 1997, p.47.

5 Rights of Successful RCFA

2

For RCFA to be successful there are many things that must be done accurately and completely. What are the key elements for achieving success in a failure analysis? What are the root causes of successful RCFA? What are the key elements that we must get "right" to nail the RCFA – the 5 rights to successful RCFA? In Figure 2.1, the five rights to successful RCFA are connected as we consider the RCFA program or process. A single RCFA can be done successfully without a certain system in place but for an RCFA process or program to be successful, the right system is also a must.

RCFA is a series process. As the saying goes, a chain is only as strong as its weakest link, and so it is with RCFA. Any portion of the chain links for RCFA that is missed will lead to error in the RCFA in the end. The earlier in the process where the error occurs the greater the miss can be. However, even if it is only missed at the last link, RCFA error is still the result.

For example, if the wrong part evidence is saved or preserved then the analysis will be executed on the wrong part. Analysis will be done for a part that didn't cause the failure. There will be some actions that may even have

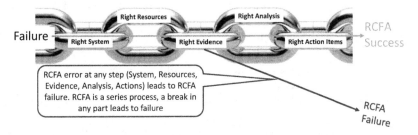

FIGURE 2.1 – 5 rights to successful RCFA

DOI: 10.1201/9781003248675-2

15

some merit but the mission of the RCFA is to solve the failure that occurred so as to prevent reoccurrence. In that regard, all that has been accomplished, assuming all the other links were done accurately, is solving a failure mode that didn't happen, that is, a random one-off Failure Modes and Effects Analysis (FMEA), to some degree.

This may be why some organizations can't seem to get any momentum on an RCFA initiative. When results are not attained after initial investment, it becomes a little more difficult to sustain the effort for future RCFAs. Successful RCFA is intentional and has a formula (5 rights). This book will summarize areas that must be in place for successful RCFA. There are other good books that discuss RCFA process methodologies such as Root Cause Analysis - Improving Performance for Bottom Line Results, Bob Latino and Ken Latino, 1999, CRC Press.

2.1 THE RIGHT SYSTEMS

The right system is not only about supporting the RCFA process but also about the overall RCFA program and the culture of an organization. The right system pieces help to solidify the organization's culture for long-term success. Organizational culture is a function of belief, behavior and recognition, as shown in the equation.

$$C = Be\left(Bh\right)^{R}$$

Where C = Organizational Culture
    Be = Belief
    Bh = Behavior
    R = Recognition

If the behavior doesn't line up with belief, then a confused and ineffective culture will result. The behavior and belief are enforced by the recognition in rewarding or correcting behaviors. A root cause failure-finding culture must be groomed and supported by having the right systems and leadership. With the right systems in place, the individuals will find success in RCFA efforts.

For successful RCFA, the entire organization must support starting at the top. At the plant level, this is the plant manager. Plants can have individual RCFA success without corporate level support if the plant manager has given support. After the plant manager, someone at the next level in the organization must own the RCFA systems. A reliability manager would be a best choice, but it may also be a maintenance manager. A reliability engineer or maintenance engineer would be some other options. At the execution level of the organization, there must be sufficient resources to execute the RCFA process and function. This sounds elementary but too many organizations start an RCFA program with good intentions but do not put fundamental resources in place for execution, which ultimately leads to failure.

For long-term RCFA success, we must make the effort sustainable and not fall victim to corporate amnesia, which breeds generational failures. That is where learnings from failures are only applied for one generation. Once that generation is gone, the organization forgets what it learned from the RCFA and repeats these failures. Too often when certain people leave the organization, the knowledge goes with them. Without good systems, these generational failures are destined to repeat. Generational failure today is more like a decade as more and more professionals change companies and locations than a few generations ago.

2.1.1 RCFA Process Infrastructure

Here are a few of the major elements in the RCFA process infrastructure needed for success.

1. RCFA Documentation: How is RCFA going to be documented? The documentation and process used not only holds the RCFA information but also guides the RCFA leaders in the process itself providing a skeleton of RCFA elements. It may be a paper-based process in a Microsoft format like Word, Excel or Powerpoint. I've seen these work, but they need to be simple and managed in the organization or chaos can result.

   Another option might be an RCFA database, either a purchased third-party system or an internal designed system. This can be an improvement over a paper-based-type format but must have a few commitments in place to be successful. First it must have an owner which is the RCFA owner – typically the Reliability Manager. Next it must also have information technology (IT) support. These systems can die and kill an RCFA program if not maintained and functioning properly. Some advantages of an RCFA database are in the ability to automate many of the functions in organizational data analysis.

   Another RCFA documentation piece is the connection of a computerized maintenance management system (CMMS). Failure data and work order coding can be an important part of supporting RCFA documentation and the database function, from the front end where the failure occurs to the back end with corrective action work orders.

2. RCFA Triggers: When is the RCFA going to be executed? There should be a clear understanding of what initiates an RCFA. Typical triggers might be a certain amount of area downtime (minutes or hours) or a certain financial impact ($) or repeating failures over time. Who will enter the initial failure information? If this person is not identified, the RCFA is handicapped right from the start. What level of RCFA will be executed? Is it a full RCFA or a mini-RCFA? A mini RCFA will typically have an obvious root cause and will not use many resources. This needs to be understood early for data collection purposes.

3. RCFA Action Items Management: How will RCFA action items be managed? Who enters this information? Where is the list? How is follow-up managed? Determining a lot of how this is executed will depend on which documentation path is chosen. Managing action items is another big advantage of having a software-based system.

4. RCFA Training: Who is trained and what kind of training do they need? General training on RCFA may be good for the entire organization but each piece of the RCFA process needs detailed specific training. The training should be on the principles, processes, and systems of RCFA. Other training areas to cover include entering information, RCFA triggers and action item management. The lead analyst may need additional training in facilitating RCFA.

5. RCFA System Health Check: As with any system, there needs to be a system check to make sure it is still functioning properly, or it could die without notice. It can be a system check on the key elements still functioning, such as that the failure events are being captured, the RCFA is being completed, that action items are being entered, etc. This can be done at a plant level or within an operating area, or both. An RCFA audit may be an audit on specific RCFAs to see if the RCFA process detail is being followed. For most organizations, this will need to be a random audit due to time constraints and available resources to execute. These audits can identify how effective RCFA's are at solving the failures. As with any system, these are living systems so, as gaps are identified, changes and action can be taken to improve as the organization matures.

2.1.2 Type of RCFA Process

One of the big questions of the right system is which type of RCFA process should be used? I don't believe there is a one-size-fits-all but there are some key principles to apply to every RCFA. We also should not force a process that is not suited for RCFA; to the hammer everything is a nail. There are also some good elements from different types of processes that can be mixed in the RCFA process to enhance and improve RCFA results. Many times, RCFA success is a matter of the probability and confidence that the root cause has been found. Evidence is not always as clear as Professor Plumb saw Mrs. Peacock take the candlestick from the library (from the board game clue).

There are many different types of RCFA methods used, such as the method of differences, cause-and-effect analysis, barrier analysis, change analysis or human performance. Most may be known by specific names such as Kepner Tregoe (KT), (method of differences process) or fish bone diagrams, the 5 Whys, and the logic tree (all being cause-and-effect analysis).

2.1.2.1 Kepner Tregoe

The Kepner Tregoe (KT) method, also called analytical trouble shooting (ATS) is a structured process using a series of questions comparing one condition to a different condition. A series of questions is asked around what, where, when, and the extent of the problem. The comparison of different conditions around what, where, and when are meant to direct the analyst to the potential root causes.

KT characteristics:

- One machine or component must have the failure while the other doesn't.
- All comparisons that are alike are ignored and considered not to be a possible cause of failure.
- Only differences identified are considered possible causes.

The difference questions can be applied to the same machine or to other similar machines or components. This can be a powerful principle to apply on some failures. If a plant has many similar machines but only one where failures are occurring on the same type of system, this is a good method to apply, especially for an analyst who may not be as expert in dealing with the failure or equipment. The differences can give clues about where to look on the system that has experienced the failure where others have not. KT can be a good supplemental tool in failure analysis but not as much as the stand-alone method.

For example, several paper machines in a plant were having dryer bearing failures. There were some machines that were not having bearing failures. While other failure analysis took place to solve the failures, the method of differences would reveal the machines not having failures had functioning desuperheater systems while those experiencing the failures did not. The superheated steam going to the dryers was cracking the inner rings on the bearings. This method was not used solely to solve the mystery but was supplemental evidence.

2.1.2.2 *The 5 Whys*

The 5 Whys is a common and very popular RCFA method used by many across manufacturing today. It is easy to learn and easy to apply to a failure, which is likely one reason for its popularity. It can be an effective tool, but it has its limitations. Personally, I like the 5 Whys for a quick troubleshooting process or a method for frontline supervision, living in the moment on the floor. Users don't typically have the knowledge, time, or resources to conduct a more thorough RCFA.

The 5 Whys is a series process where the question of "why?" is asked 5 times. That is all there is to it. Now you are trained in the 5 Whys. It can be an effective and fast troubleshooting thought process. It is a linear process and does not involve branches or broad lists of possible causes. It is strictly dependent on the analyst's experience and knowledge to identify the correct 5 Whys so as to get to the root cause.

For example, a stock tank is experiencing poor mixing showing up in the process. Troubleshooting and applying the 5 Whys might look like the progression below.

Why 1 – Why poor mixing? poor agitation
Why 2 – Why poor agitation? Agitator not turning
Why 3 – Why not turning? Belts burned off
Why 4 – Why belts burned off? Excessive load
Why 5 – Why excessive load? Buildup inside tank

On the surface it appears like this has been a successful analysis; however, lets evaluate further. We will even assume that all the findings at each question were validated as to be accurate. Where might assumptions be made? After the first question. Poor agitation may be true, but the cause may have causes other than agitator functional failure. What if one of the process streams controlling consistency had failed?

Because the buildup was found in the tank, it seemed to make sense. Maybe at the very least, the tank buildup may have been a contributor to the failure, but it may not have been the root cause. What are some other possible causes? The last time maintenance was done on the agitator the spare had the wrong sheaves, which left the belt drive with the wrong # of belts and turning the wrong speed, thus causing the belts to be insufficient for the drive load.

Other possibilities for the belts being burned off or not lasting are misalignment, incorrect tension, motor too large for application (which could have happened on the last spare changeout). The 5 Whys is a linear thought process but does not leave room to capture other possible causes of that failure mode.

One other element is that all the roots may not be solved after the 5 Whys. For instance, in this case why did the tank have buildup? If agitator failure is the reason, then tank buildup can't be the root cause of agitator failure. So the failure process isn't in the proper order. In the example the tank buildup is identified as the root cause so what caused the tank buildup? Was there a process upset that overwhelmed the system? Were there consistency control issues or something else?

For an organization beginning in RCFA, the 5 Whys is probably a good place to start, but if better probability or return on the RCFA investment is desired, a better method is probably needed.

2.1.2.3 Failure Mode Based

This is a cause-and-effect type of methodology where the basic failure mode is defined, and then possible causes are linked to each other, branching out in multiple directions so that no possible causes are missed. Only the possible causes are listed which can cause the actual failure mode that occurred on that failure. An example of this can be shown in Figure 2.2 below. This is not a complete failure tree but an example of how it is structured.

While there are advantages and disadvantages of every type of process, it is important to remember that the process itself will not solve failures. If you have the right answer to the problem, it really doesn't matter which process you use. However, having the wrong answers to the problem using any process will not result in the correct solution. People solve failures; the methodology or process is just a tool to find the root causes. The people doing the analysis must do all the work in searching out all the possible causes.

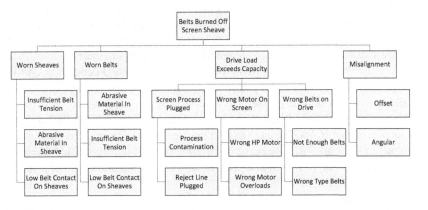

FIGURE 2.2 – Failure mode cause & effect logic tree

2.2 RIGHT RESOURCES

Since people solve failures, the biggest resource in successful RCFA is the people. In a Dilbert cartoon, the boss asked Dilbert if he found the reason for the failed project. His answer was "the problem was people." The boss asked, "the wrong ones?" Dilbert answered "Don't over-think it." It takes engaged, motivated, and experienced people to solve the most difficult failures. It takes people from all parts of the organization, from engineering, operations, maintenance, purchasing, and management. It also takes some external people to participate to get to the root causes.

It has been shown in organizations that the best success comes when the right systems and the right people come together. Poor systems and the right people will only frustrate the people and good systems with the wrong people will also not produce the most successful results.

It is important to have enough resources to carry the RCFA from beginning to completion. This is the only way to build credibility in the RCFA program and yield the real results to the organization.

2.2.1 People

One of the key groups of people needed for successful RCFA is reliability engineers. They are typically the lead analysts on most RCFA. There should be both mechanical and electrical reliability engineers for a comprehensive RCFA approach. Typically, a reliability engineer can support one to three operating areas. The size of the area and how many failures an area has will determine how many reliability engineers are needed. For a focused improvement in an area, a dedicated reliability engineer or engineers is required. More mature organizations can operate with fewer reliability engineers.

Other subject matter experts (SMEs) are also needed depending on the type of problem. It may be a pump system problem or a metallurgical or electrical drive or controls problem. Whatever the type of problem, a SME will be needed to answer the hard questions when it comes time for the analysis portion of the RCFA. We can't expect inexperienced or untrained employees to fill in what an SME can bring to the table in solving complicated failures. This may be one of the biggest reasons why many companies do not achieve successful RCFA.

Another key element about the people who execute RCFA is that these leaders take time to develop. No matter how much is done to shortcut it, experience is one thing that can't be short cut. Experience can only be gained in time. Therefore, it is important to invest in young professionals as the next generation of reliability engineers and professionals come along. This will ensure success will be sustained in future generations.

There are many other people resources needed outside the plant, which include quality precision rebuild shops to participate in the RCFA process. They are key in every failure that is not addressed at the plant level. There are external services and labs that are also important to the RCFA process, including oil analysis, vibration analysis, and metallurgical services.

2.2.2 Budget

Successful RCFA requires commitment by the organization with respect to budgets. There must be a budget for resources to execute RCFA. That also includes training for all people involved in RCFA. It also may require budgets for travel to receive training or to be at a rebuild shop to witness teardown after a failure. There must also be budgets to support these internal and external activities of investigating failures, which may include additional analysis by technicians.

If a plant does not have reliability engineers or professionals to lead and execute RCFA activities, then a justification must be made to add them. A typical reliability engineer return I have heard is ~10:1 minimum ratio. Meaning if it cost $100K a year in salary, the cost avoidance or return savings of a reliability engineer would be $1MM. I've seen this many times to be as high as $2MM. A similar estimate can also be said of other reliability professionals such as vibration analysts.

2.2.3 Tools

There will be times when extra precision tools are necessary in performing RCFA or executing action items to prevent future failures. These may include all kinds of precision tools for measuring machines, components, and parts. It also may include extra investigative tools for the RCFA analyst such as special magnification tools, cameras, hardness testers (metal and elastomers), surface roughness testers, flow or velocity meters, power measurements, and the list goes on.

Successful RCFA is messy and requires the people, budget, and tools to execute. These things are not in place overnight but build gradually over time. It has been said that the journey of a thousand miles begins with the first step. Start building your resources to enhance RCFA execution.

2.3 RIGHT EVIDENCE

The RCFA process itself begins with the right evidence. The evidence begins with the failure event itself. As much information as possible must be collected about the event. This includes the time up to the event, during the event and even after the event for clues as to how the system was operating around the failure. Eyewitness interviews as soon as possible after the failure can be critical. There have been times when the person standing in front of the equipment, watching the failure, was a key factor in determining the root cause. The promptness of these interviews can be important as little details get forgotten quickly. People also tend to, sometimes unintentionally, bend the story to fit a narrative that they believe is the cause.

2.3.1 Event Investigation

Whenever possible it is always best to interview all the people involved in the failure event as soon as possible. This is also where having the right systems can help capture key details of the failure event such as with rotating shifts where it may not be feasible to have firsthand interactions. Capture any process trends or equipment performance conditions before and after the failure. Sometimes these subtle things can mean the difference in RCFA confidence.

I had a felt roll failure once where the roll ended up tearing out of the machine and destroying parts of the machine as it found its way to the basement. Many different people were anxious about the cause of the failure. The basement of every machine was barricaded off until the failure investigation was complete and corrective actions completed. We really needed to find the root cause so that normal operation could resume like checking lube systems under the machine. Some were concerned about the machine frame fracture being the root cause. Others thought the bolts holding roll vibrated out. One key piece of the investigation revealed that there was one operator who was an eyewitness to the failure. He saw the roll come to a sudden stop and when the felt stopped slipping on the roll it grabbed the roll and ripped it out of the machine. The frame wasn't previously cracked and bolts were not missing. There was other evidence from the failed parts that proved this as well, but the failure event testimony helped to steer the investigation quickly away from wrong possible causes. The root cause of this failure was the bearing locked up when the lube line broke feeding the bearing. Why did the lube line break? The roll had housing fits that were excessive which allowed vibration which fractured the lube piping.

2.3.2 Failure Scene

Failure scene documentation comes before full event investigation because of how rapidly key evidence can be removed, altered, or erased. There is only one small window of opportunity for gathering failure scene evidence. Once the part is moved, cleaned up, evidence washed down the drain or disassembly occurs, that key evidence is lost. This evidence can make some seemingly difficult RCFA rather simple. Unless the failure is chronic, the analyst only gets one shot at gathering the evidence and solving.

Take lots of pictures. Digital pictures are cheap and easy, unlike the old days when I remember using a Polaroid to take pictures. You never can see all the details or facts from a failure at the time of failure. Many times, the people involved in performing root cause analysis are also involved in the reassembly. The repair is usually more of the focus at the scene so the pictures may be the only insight into the failure scene itself. Reusing parts from a failure for repair also stresses the importance of having photos of the failure scene.

As far as the component failure analysis is concerned, it starts with the part. The right evidence is the actual failed part. Without the failed parts, the RCFA process becomes a guessing game. The failed parts should also be preserved, as much as possible, as found. Don't clean up the failed part, removing evidence of how the failure occurred or conditions surrounding the failure. This is especially true of components where lubrication is in play. The condition of the lubricant, the amount remaining, can be critical to determining the root cause of failure.

If possible, always try to witness disassembly at the time of failure and the components at the shop. There can be much to be learned by seeing how things look as they come apart, which someone not focused on solving the failure may not notice. Figure 2.3 shows two bearing housings and the inner race

Drive End (Failed End)   Opposite Drive End (Non-Failed End)

Same grease cycle as ODE, little grease amount in failed DE bearing

No grease reaching bearing as housing wasn't packed

FIGURE 2.3 – Motor bearing housings

from motor bearing assemblies. The drive end motor bearing catastrophically failed. The motor had been operating for about 9 months. The motor bearings had been greased once, which was not enough, but it would not have mattered. The motor bearing housings were not packed with grease. The opposite drive end (ODE) housing shows where the fresh grease came into the housing but just sat in the empty housing. The same thing happened on the drive end (DE). The DE bearing failed first as it had the most load and a higher operating temperature. This is also a good example of using the simple method of differences, comparing similar bearings on the same equipment to determine the root cause. While what was different was the bearing load, both bearings were lubricated at the same time so the bearing that didn't fail confirmed how little grease was in both. The DE bearing had no lube residue but the ODE bearing confirmed the lube condition.

One final word about the failure scene – use caution in dismantling from failures, especially catastrophic failures. The catastrophic failure scene can leave many unsafe conditions. It may be necessary to involve engineering or guys that are rigging specialists to slowly pick through the pieces. I remember a game we used to play when I was a kid called "pick up stix." The sticks were all dumped in a pile and you had to try and pick up as many sticks as you could without making another one move. That same type of approach is how you want to deal with a catastrophic failure scene. There are sometimes broken structural supports, so evaluate loads at each step of disassembly.

2.3.3 Failure Type

One of the first things to do at the failure scene is to identify the type of failure. Is the failure a single failure or multiple failures? Multiple failure typically is catastrophic in that more than one component has failed. If the failure has an obvious single failure, then concentrate around preserving that part and joining parts for analysis.

For multiple failures, keep all failed parts until determination can be made to identify the primary failure and secondary failures. Primary failure is the component failure which precipitated the functional failure of the asset. The primary failure is also the root cause of the secondary failures. The focus of the RCFA should be on the primary failure. This can be a challenging process on catastrophic failures as, with the chicken and egg scenario, these can be hard to separate. It is easy to get distracted by the severity of secondary failures. One set of questions to ask to determine primary and secondary failures is to check each failure as a possible cause of the other. In other words, can failure A cause failure B or failure C and so on?

Determination of primary and secondary failures is about determining what the right evidence is or which is the right component to do more analysis on. An apron feed conveyor experienced a chronic linkage failure. The belief was that there was an issue with the chain link material and it was wearing thin until the pin ripped through and caused a functional failure of the conveyor. This was partly true, but more to do with what the primary failure component causing this was.

The chain links were reassembled as they were at the time of failure as shown in Figure 2.4. Looking at the failed pieces, here are some of the different items noted on the multiple failures of the catastrophic conveyor failure.

Failure A – Chain link worn and hole torn out
Failure B – Chain bearing failure
Failure C – Conveyor slat bent
Failure D – Chain links bent

After the scene investigation, it was obvious that failures C and D were secondary failures after only one side of the conveyor chain came apart and bent those components as a result. That left failures A and B. Comparing failures A to B, can failure A, the chain link being worn, cause failure B, the bearing experiencing catastrophic failure? No. Then failure A can't be the primary failure. Can failure B, the bearing failure, cause failure A, the chain link worn and ripped out of the end? Yes. Again, confirms that failure B with the bearing failure is where the analysis should be focused.

If the analysis focused on the wrong failure – the chain link – hopefully the analysis would eventually end up back at the bearing, but that is no guarantee. In this example, there was already discussion about changing chain link material. So, not getting the right primary failure can have all kinds of negative consequences in the RCFA process.

Look again at the picture in Figure 2.4 and note that the opposite end of the chain link that failed, which has the notched hole that fits on the bushing, is in like new condition. Yet, that assembly sees the same conditions as the opposite end that pulled out. How can this be? Well, the failure isn't due to anything to do with the chain material because that is the same across the same part (the applied method of differences). A spray lube system sprays on the chain as it rotates.

Even though we are not really doing failure analysis at the component level yet, there is quite a bit of analysis in gathering the right evidence. So how else can we conclude that the bearing failure causes the linkage wear and tear out? We must look at the detailed chain design and how the pieces fit together. In Figure 2.5, the wheels of the apron chain assembly have antifriction roller bearings. The center pin and bushing only pivot as the chain

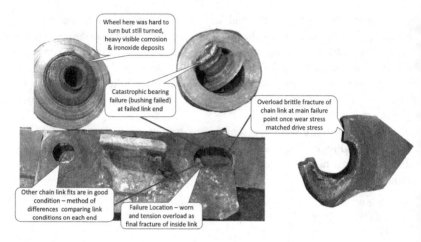

FIGURE 2.4 – Failed apron conveyor chain

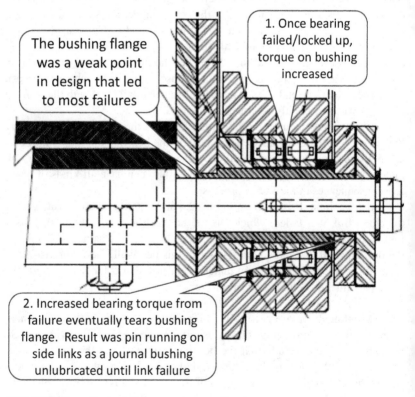

FIGURE 2.5 – Chain design

goes around the drive and tail sprockets and remain static the rest of the time during conveyor operation. The chain side links are fixed on the pin (for the outer link) and fixed on the bushing (for the inside link) so they really are not rotating bearing points for normal operation. The antifriction bearing takes the rotating motion.

From Figure 2.4, the end of the chain link with the ripped-out end was the end with the failed bearing. When the bearing failed (locked up) the bushing was destroyed which also pulled out of the link. Now the chain has the pin running as a journal bearing on the chain links, and the chain links wear until thin enough that chain tension tears through the thin remaining link.

So, now with the proper evidence the failure analysis can commence on the root cause of failure, focused on the bearing failure. Without a lot of explanation, insufficient lubrication was found to be the root cause of bearing failure. The lube oil spray system was found not functioning for a long period of time. Also note the bearing near where this failure occurred, as shown in Figure 2.4. It had a lot of corrosion (iron oxide present). The bearing still rotated (had not experienced functional failure Level 4) but had experienced a Level 2 or 3 failure.

2.4 RIGHT ANALYSIS

2.4.1 Failure Mode

Once the right evidence is obtained, the right analysis starts with the failure mode. The failure mode can be defined as any event which causes a functional failure.[1] This definition is relative to the FMEA process regarding equipment functions, functional failure and failure modes. It must be expanded and more clearly defined to fully understand how the failure occurred. How the failure mode is described will immediately determine the success of the RCFA. It will either lead toward the root causes or lead away from the root causes. One doesn't have to be expert in describing the failure mode, but it must be enough for a subject matter expert to complete the analysis.

An example is a shaft breaking. An elementary description by someone not experienced in shaft failure analysis might just describe the failure mode as "shaft fracture." A better description might be "shaft fracture at the shaft step behind the bearing." A more advanced description by an SME might be "high cycle, low stress rotating bending fatigue with severe stress concentration."

The elementary description of the failure mode is fine if it is followed up by the right SME to finish the description of the shaft fracture. Otherwise, the RCFA is just guessing at what the root cause of the shaft fracture is. With the best descriptive definition of the failure mode, the RCFA can focus on solving that specific failure mode. RCFA is a focused effort to solve only the specific failure mode that occurred. FMEA, is a broader approach to look at all failure modes to design systems to manage all possible risks.

The more detailed the failure mode description the more focused the RCFA can be. If the failure mode is not developed well enough, then seek out an SME to develop the failure mode. The failure mode describes what you see on the failed part. This becomes the road map for the way to successful RCFA. Whichever type of RCFA process is chosen or used, it is not a make-or-break choice but staying true to the cause being linked to the correct failure mode is. Once the correct failure mode is defined, it is time to establish possible causes for that failure mode.

2.4.2 Possible Causes

Possible causes are for the correct identified failure mode to address. In RCFA, we are only solving the failure that has occurred, not the potential failures which can occur. Consequently, the root causes can't be found if the possible causes are not identified. Engaging the right SME in cross functional areas from process and equipment to plant operations will be required to come up with the right possible causes. There is no need to waste time on a possible cause that can't really cause the failure mode of that failure.

Proving and disproving possible causes is where the messy work really comes in with RCFA. Usually, the failure mode and even possible causes may come quickly but finding the right way to investigate and prove and disprove possible causes is now the challenge.

Another delicate part of the analysis portion for possible causes is determining how good the data is in proving the hypothesis. Opinions and gut feelings are not truth. It has been said, trust God but all others bring data. Truth can be absolute truths from various sources such as proven science or learned science through observations and experiences, but you can't cheat physics.

There are so many places to access data for analyzing the possible causes of failure. If we are looking at the physical asset, then we need to look in areas where the physical asset has some hard data around the condition. This would include maintenance work history, procedures, rebuild reports, preventative maintenance (PM), and predictive maintenance reports (vibration analysis, oil analysis, etc.).

Additional data typically needed is around the design baseline for the equipment such as drawings and operating and maintenance manuals. There is a need to understand the design theory and operation of the pieces being analyzed. What are the key design variables that dictate performance, capacity, and design such as load, pressure, speed, temperature, flow, volume, power, torque, etc.? A big part of the analysis may be in looking at these key variables to see if there are any exceptions and how severe they were relative to the root cause of failure. These are your inputs for analyzing the physics of failure.

2.4.3 Physics of Failure

Applying the physics of failure (POF) is one way to identify and keep focused on the possible root causes. All our machines have elements of physics involved in the failures. Knowing these principles of engineering and physics to the failure can serve as a scientific guide to solving. We will show several examples in other chapters of the case studies of this principle used in RCFA. Many times it will involve the formula to some mechanism and other times just some application of fundamental laws of physics. In the end, the failures that happen can't cheat physics, even though some may try.

Consider a coupling hub that has slipped on a shaft. What could cause that failure mode? Knowing the drive torque, what variables should be investigated? Apply the physics of failure to the failure mode and find some possible causes from the formula below.[2]

$$T = Eitlf\pi$$

Where T = torque fit will transmit, lb-in
      E = Modulus of elasticity, psi
      i = interference fit, inches
      t = thickness of hub, inches
      l = length of hub, inches
      f = coefficient of friction

From the physics of the application, the possible causes are given as

1. Incorrect materials used (E)
2. Incorrect hub fit (i)
3. Incorrect hub thickness (t)
4. Incorrect hub length (l)
5. Incorrect friction between hub and shaft (f)

These are some possible causes to start with (assuming the driven torque did not exceed the design). There are likely many more details involved in the RCFA but POF is just another tool to add to the process. POF requires some engineering expertise so adding these resources to the RCFA team is critical.

2.4.4 Failure Patterns

When a failure occurs, it is many times helpful to identify the failure pattern of the failure. Six failure patterns can give clues for what to look for, as failure analysis is being done on the failed part, and can help with action items. The six failure patterns are shown in Figure 2.6 and are gleaned from the reliability centered maintenance (RCM) principles. While the original study came from the airline industry, it has been repeated in different fields with some different numerical results. While the percentages may vary from industry to industry, it is what each pattern reveals that is important. There are three age-related failure patterns and three random failure patterns.

All these patterns have time as a common factor so we must first understand how long this part should last if achieving full life. For example, full life for a V-belt might be 3 years while full life on a spherical bearing running on oil might be 25–40 years. A contact lip seal may reach full life at 6–12 months while a labryth seal may last a lifetime or 50 years. This can help us to understand whether the component failed due to wear out at end of life or because of infant mortality.

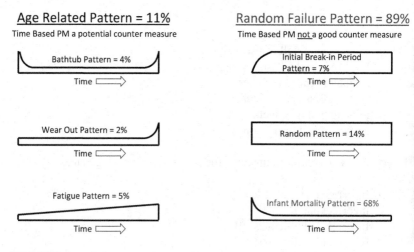

FIGURE 2.6 – Failure patterns

The bathtub curve has elements of wear out and infant mortality combined. Wear out may be on a part such as a brake pad or clutch on a vehicle where, after so much time or so many cycles, it is worn out. The distribution of failures here happens over a certain amount of time (cycles). If a part has reached its wear-out life, then the failure may not have been due to anything but just time. The surprise of the failure is that it may show the need to improve some maintenance systems to allow better planning, but that the component had lived its life.

For infant mortality, the failure occurs early in the component's or asset's life; hence, the term infant. Infant mortality can point to several areas where failure can be introduced to the machine. Some of these might be the commissioning and start-up of the equipment from recent replacement or repair. These are areas where human error on these activities can be the source if the root cause. With infant mortality being the highest in most of the failure pattern studies, it is always a factor to consider. A heightened awareness is good to have for the infant mortality period, from starting-up machinery, and especially to recently worked-on machines. This is the time where probability of failure is the highest.

A roll gearbox was recently rebuilt, and the roll was installed in the machine. A catastrophic failure resulted during start-up as the gearbox had a sudden thermal event within the first 15–30 minutes after start-up. The findings were that the new labryth seal was not installed correctly and slipped on the shaft and galled or locked up on the mating labryth seal. There are many times when I would rather have a machine that has run beyond its infant mortality period to know that its design and assembly has been proven. I took a flight once on an airplane that still had that new car smell. I was instantly more nervous than I normally would be. The stewardess told us to relax for the flight, since this was a new aircraft that had only made a few flights. For a few of us, it didn't quite have the effect she was after.

2.4.5 Ending Analysis

When conducting the failure analysis there comes a point where the question of when to stop the analysis needs to be asked. A quick answer is to stop at the level of feasible control or correction. If the analysis goes beyond this the probability of prevention is either low or not financially feasible. The belief that an RCFA is not successful unless it prevents the failure from ever occurring again is not feasible in almost every case due to the level of control from the root cause as well as the unrealistic financial cost to execute such a demand.

For example, the root cause of a bearing failure was incorrect bearing installation when the rebuild shop removed too much radial internal clearance (RIC) from the bearing. The RCFA can ask for rebuild reports, and even make suggestions for modifying procedures for the shop, but the control of that process is not in the plant's control. It can happen initially but let plant personnel change over time and those fine details and learnings can be lost even if procedures are in place. It is likely not feasible to send plant quality control to inspect every rebuild. In the case of the bearing installation, it is also not feasible to inspect an item once it comes to the plant. Having identified the root cause of failure and sharing the learnings must be considered a success for an RCFA such as this.

At the end of the RCFA, I have sometimes found it good as a sanity check to rethink the series of events in the progression of the failure and see if it makes sense when all the facts are put together. It is easy to compartmentalize each of the analysis steps and findings and lose track of their connection. If something doesn't add up, then there may be something missing in the analysis. This is not part of the analysis but just a way to put the life cycle of the failed equipment together and see if the whole seems feasible. Individual pieces will many times make sense but putting the pieces together sometimes can reveal holes in the hypothesis.

When a definitive root cause is not found or known it can seem a mystery but there is no such thing as a mystery, only a lack of understanding. Something has either been missed in the investigation or the problem is just more complicated than the resources available can solve. This probably happens often on many plant failures especially chronic failures. After all this is how they become chronic, they have not been solved, they just repeat.

Also, many RCFAs have some element of uncertainty that the root cause has been found. Where there is not a clear definitive root cause, what do we do? How do we determine a root cause or maybe more importantly how do we proceed? There are a couple of ways to approach this.

Perhaps the best way is to compile all the evidence and take your best stab at the root cause. There is always much circumstantial evidence just like a court case. It is many times the cumulation of evidence that confirms the potential root causes. Often, there just isn't hard data to confirm a hypothesis or possible cause but there are several aligning datapoints that support the suspected root cause. If the probability on the root cause is low, you don't want to invest a lot of resources into an action item that may not address the root cause. Taking this first approach is a way of doing something based on initial data but the door is always open to revisit the failure again.

When the failure cause can't be determined the action items may become more investigative or even extreme so as to make the problem go away. That may involve working around the root cause of failure even though a root cause was never really found. It may involve a redesign and over-design to ensure failure is avoided in the future.

A pickup roll tending side bearing was experiencing a Level 2 failure within days and weeks of installation and running-in the machine. Extreme vibration was evident on the roll, but it continued to operate. This same bearing and roll were used on four rolls on two different machines. Analysis went on for over 12 months involving every bearing expert and third-party lab we could engage. The wear pattern on the bearing roller was like nothing ever witnessed by every bearing expert. The rollers had circumferential grooving smooth across the rollers. Every installation and mounting condition as well as bearing material types of issues were analyzed, and nothing could be found. Finally, utilizing the method of differences in the machine rolls, it was observed that there was one roll that had no such failures. The only difference was that a different bearing manufacturer was used on that one, thus the only thing we knew we could do was install that brand. Once we did, all failures were gone, and we never had any idea why or what. We continued to use the brand bearing with the failures on other applications with no issues. We never got a root cause of failure, but we did make the failures go away.

2.5 RIGHT CORRECTIVE ACTIONS

In the chain of successful RCFA, all the previous stages of the RCFA process can be executed to perfection but if this last step is not executed correctly then there is not a return on the RCFA investment. It is like leading the entire game but then missing the final tackle that would have won the game. The final score says you lose on that last play even though you may have played the game well enough to win. I've seen many in RCFA expend a lot of energy on finding the right root cause, yet the right action items are incomplete. The root cause is reported out and everyone is happy and ready to move on to the next failure.

How do we ensure that we get the right action items completed so as to have a successful RCFA? One of the first ways is to have action items as part of the right system. There must be a system to record, track and review the progress on action items. The format that it is in doesn't really matter but organizational discipline does. The action items can be tracked in a spreadsheet, on a meeting-room whiteboard, or in an RCFA database. Some database systems today may be set up to allow communication to CMMS or email to track progress, work orders and those assigned.

A regular review of RCFA and action items should be done by the appropriate areas at least monthly. Realistic completion dates should be set so that adjustments can be observed when things are slipping. Putting completion dates that are too soon or too far out really doesn't accomplish much.

2.5.1 Type of Action Items

To ensure that no action items are missed, it is good to identify your action items. Most of the time you will want to address the physical, human, and latent causes in the action items. The physical root cause action items are always usually addressed. Everyone knows the physical component that needs to have the roots addressed. It is the human and latent roots that are likely to be overlooked. The reasons are partly because they are the hardest to address and the most sensitive with regards to people in the organization.

Human root cause action items will need to be handled carefully. There should be no discipline-type action items from RCFA on human root causes. This is the fastest way to kill an RCFA culture where the organization wants to engage in the RCFA process. Human root action items can focus on ways to improve human interaction around the failure to prevent human error. It may be a procedure or engineering-out human error on the machine, but these address the human root causes.

The latent root causes are also challenging as they branch out further into the organization beyond the operating area or plant control. The latent root causes will really test the character of an organization. If leadership does not support latent root cause action items, then this part of the action items will have to be re-evaluated. The level of where to stop will be shallower, corresponding to the level of support in the organization.

For example, if one of the latent root causes of failure was insufficient training for maintenance on installing a component, then the action item may be to put a training program together for that. If the organization will not support executing that training program due to budget or other resource constraints, then a different approach must be taken. Maybe the action required is for training to be done at a local level in that area. In that case, that may be all that can be executed successfully so there is no need to have a corporate level action item. Stop at the level of feasible control.

It is also good to refer to failure patterns (see Figure 2.6) as an aid for covering all the action items. If the failure is infant mortality and some equipment was rebuilt incorrectly then procedures and training are certainly possible action items that will likely need to be considered. If the failure pattern is a time-related wear-out zone, then a preventative maintenance task in some form may need to be considered.

Random failures will eliminate PM for the most part since it is not time dependent. Many times, after a failure, the first reactive action item is to add

PM in the system to check that item. It usually has no detail of what to check or why it is checked but it does make the stakeholders feel safe, knowing that someone is looking at their problem child. This is what I call PM creep. After a while, the system is full of PM and all the resources are used up on non-value-added PM. I've seen studies where as much as 40–50% of PM adds no value. My experience in plants is that this is pretty accurate. PM does consume a great number of resources.

One time, a doctor bracket cracked on a machine on which the machine manager wanted a PM inspection of the new bracket at every outage. What was the inspection? What was it preventing? The maintenance tech was just walking up and looking at it. There was no value in this at all. A fatigue-type failure is not addressed by PM.

Here are a few items for checking in your PM before adding more from an RCFA. Consider these to avoid PM creep.

- Properly define the failure mode you are wanting to prevent with PM. Also define the failure pattern for that failure mode and P-F interval. This will set your PM frequency to be half of the P-F interval.
- If you do not find a problem with PM every 5–6 times it is executed there is something wrong with the PM – wrong detail, wrong frequency, wrong failure mode, performing PM incorrectly or just pencil-whipping?
- About 15% of your work should be driven from PM work. If you are not identifying corrective work from PM then something is wrong.

There is another philosophy that can be helpful when considering possible action items. As with any failure, there are different measures of insurance to prevent failure at different levels. The Swiss cheese model is one that, for failure to happen, the holes of different systems must line up.

An example here would be a failure where we had just changed a pump rotating assembly. The pump was started up and failed very quickly, was overheating and the bearings failed. The RCFA was straightforward when the pump was found to not have any oil in it. There are many levels or layers where this failure could have been prevented, but it wasn't. These are areas where the action items can address deficiencies in different systems, as shown in Figure 2.7. The failure could have been prevented at each level but the holes in each activity eventually led to failure. Planning and job preparation should have made sure that the oil was available in the job materials. The pump didn't even have a sight glass mounted on the housing. If the installation check sheet had been followed properly, the pump would have been lubricated. For basic care rounds, the pump ran for a few

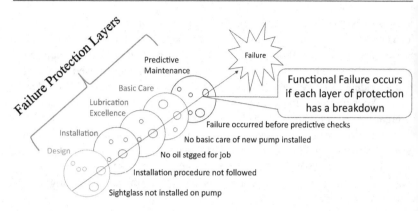

FIGURE 2.7 – Swiss cheese model

days. Had the operators gone and looked at the new pump just installed they would have seen that it was running very hot. This is another example of infant mortality. One of the highest risk times of an asset life is at start-up and shortly thereafter. That makes it a good time to do some condition monitoring.

2.5.2 Execution of Action Items

It is good to list all possible action items, even the ones that are not going to be executed. Depending on the confidence of the RCFA root cause findings there may be different decisions about which action items execute. Factors such as the impact of the failure and the desire of the organization to resolve can determine which action items are required. There are always various levels to addressing the failures, each with different levels of effectiveness and cost. Easier, faster, and cheaper actions will typically be executed first but if not successful the other more invasive ones may be used later. It is sometimes much like the progressive steps of diagnosis or treatment experienced when you go to the doctor. You don't immediately go straight to surgery, but the less invasive treatments are completed first.

One way to break down these action items is in the short term and the long term. The short term can be executed in a relatively short time and with a many times smaller budget. Some short-term items can be completed immediately or within the week. Long-term action items are those that may require a special project or approvals which often involves higher spending or capital expenditures. While long-term possible action items may remain in the root cause main document, the long-term items are typically transferred to another system for completion such as work order backlog or capital project lists.

Another type of action item is an investigative action item. These can be items that need to be done as part of the analysis or follow-up items from the RCFA. More complicated RCFAs may have many investigative action items to address before finally being able to compete the analysis.

When assigning action items, understanding the resources available is important. When action items are assigned to the wrong resources, for whatever reason, they likely will not be executed correctly or even at all. Lack of resources may even restrict how many items are entered for action. Better to complete a few correctly than halfway complete a lot.

2.5.3 Closing the Loop

One action item that is good to add if possible is to follow-up once all other action items are complete and see if the desired outcome has resulted. Did we find the root causes? Did the corrections deliver the right results to prevent future failure? If resonance vibration was the root cause of failure, then after all the corrective action items are completed, we need to check the system to see if resonance is still present.

Sometimes it may not be easy to check certain deliverables to know, but there are usually some indicators that can be verified either by looking at a measurement or a maintenance follow-up to verify the condition. It may need to be checked during the next outage or equipment rebuild to see if the previously failed part has the failure mode returning, such as corrosion or pitting.

There may be cases where the follow-up is to just wait and see if the failure repeats. The important thing here to note is to still be observing conditions along the way to look for new information that may help get to the root cause of the failure that the first pass corrections may not have addressed.

NOTES

1 RCM II, 2nd ed John Moubray
2 Ref. Standard Handbook of Engineering Calculations, Tyler Hicks, 3rd edition, p.3.117

Key Factors in Fatigue Failure

3

3.1 FUNDAMENTALS OF FATIGUE FAILURE

Fatigue is a relatively new subject of study, and has only emerged within the past 200 years. In the early 1800s a German mining administrator, Wilhelm Albert, observed how some mining chains would break after small, repeated loads even though they were not overloaded. He did some testing on cyclic loading on chains and in a paper he published in 1837, determined that the number of cycles was a major variable in fatigue failures. This marked the beginning of the study of mechanical fatigue.

Fatigue is a common failure mode where cyclic (alternating) and repeating stresses on a part cause the part to fracture. Fatigue stress is a stress below the ultimate tensile strength and most of the time below the yield strength of the material. That failure stress is typically low is a real mystery to most. The example given many times is that of bending a coat hanger but in that case the hanger material is stressed beyond yield to permanent set after the excessive stress. The coat hanger stress is only repeated 3–4 times and fracture occurs.

Fatigue stresses are typically repeated millions of times. For example, a shaft operating at 1800rpm with bending stress will experience tension and compression cyclic stress 1800 times a minute. The shaft with a bending stress has the top half of the shaft in tension and the bottom half in compression. In just one year that would be 946,080,000 stress cycles of tension and compression on the shaft. Of course, a shaft will operate with many types of stresses, some of which are constant (non-cyclic) such as drive torque and some that are cyclic such as bending. It is the cyclic stress which causes fatigue failure.

Since the cyclic stress which causes fatigue failure is typically below yield strength, the initiation and progression of fatigue cracks are almost always hidden until the final fracture. This is what makes fatigue so damaging and dangerous. A static overload failure will stress a part beyond yield and a bent

DOI: 10.1201/9781003248675-3

part will be a result. This can be visually observed either by looking at the part or observing the operation of the machine, as vibration will reveal a bent shaft or other defect on a rotating part.

Fatigue failure begins with a small crack not visible to the human eye and which is even difficult to find with advanced non-destructive testing (NDT). During the growth of the crack, it has periods where it grows and where it doesn't. Progression marks are where fatigue crack growth does not occur. Initially the crack growth may be slow and progression marks are visible. As the crack propagates, it likely will speed up and progression marks may disappear. The fatigue crack progression will continue until stress on the part exceeds the remaining stress area at which point an overload fracture will occur. This area is called the final fracture or fast fracture zone. So, the fracture surface has two distinct zones – a fatigue zone and a final fracture zone. There are many things to be learned about the fracture surface, about which metallurgical services can assist in providing key information. There are also many books available on fractography that provide help in the field when looking at fracture surfaces, which are beyond the scope of this book. However, there are the mechanical aspects of fatigue that I do want to cover with respect to RCFA where these principles will improve fatigue failure analysis.

All components and machines operate with some cyclic stress, either from machine vibration or rotational forces, but the question is why do only a small number of machines and components experience fatigue failure? Most parts operate with infinite fatigue life. This can be understood beginning with the S-N curve.

In the mid-1800s, railcar axles began to have failures from fatigue where significant events were taking place. A German railroad engineer, August Wohler, developed a rotating bending fatigue test on railway axles to help solve these failures. The results of his early work led to the classic S-N fatigue curve (sometimes called the Wohler curve). The S-N curve is a plot of the material fatigue strength versus the number of cycles which can be seen and explained in the case study in Figure 3.4. Fatigue can be defined as low cycle fatigue ($<10^3$ cycles) or high cycle fatigue ($>10^7$ cycles). The infinite region of the S-N curve is known as the endurance limit of the material. A simple estimate for the endurance limit of steel material to ultimate tensile strength is shown below.[1]

$Se' = .504Sut$

Where Se' = endurance limit for steel <200kpsi Sut, psi
   Sut = ultimate tensile strength of steel, psi

The fatigue strength of the material at a set number of cycles from the S-N curve can be expressed in the relationship below.

$Sf = aN\wedge b$

$$a = (.9Sut)^2 / Se$$

$$b = -\frac{1}{3}\log\left(\frac{.9Sut}{Se}\right)$$

Where Sf = fatigue strength, psi
    Sut = ultimate tensile strength, psi
    Se = endurance limit, can also use Se', psi

If the cyclic stress is known (substituting σ for Sf) then the expression below may estimate the number of cycles for fatigue failure.

$$N = \left(\frac{\sigma_a}{a}\right)^{1/b}$$

Where N = number of cycles
    σ_a = alternating stress, psi

The above endurance limit is the endurance limit of the test specimen, Se', but in the real world there are factors that reduce this endurance limit. These factors can be the mechanisms that reduce the endurance limit to a point where the cyclic stress is below infinite life. In other words, this is the point at which the cyclic stress > endurance limit with a finite number of cycles exists.

So, what are some of the reduction factors which can drive down the endurance limit? The endurance limit factors are shown below.

Se = Se'KaKbKcKdKe

Where Se = endurance limit of mechanical element (part), psi
    Se' = endurance limit of test specimen, psi
    Ka = surface factor
    Kb = size factor
    Kc = load factor
    Kd = temperature factor
    Ke = miscellaneous effects factor

The surface factor, Ka, is a factor determined by the surface finish of the material. This can range from 1 (no reduction) for a highly polished mirror finish to.4 with some forged material or with a rough surface finish. A surface finish of 250Ra may have a surface factor of.75 where, if the surface was finished to a 32Ra, it may be .95. It may not sound like much but when applied with a high cyclic stress, it can be the difference in infinite life and fatigue failure. There are many causes of surface finish defects such as part design, machining operations, machining speeds, alloy used (machinability), machine tool condition and machinist experience, just to name a few.

The surface factor can be such a huge variable affecting fatigue strength. In bending stress, the surface stress is where the stress is the highest. Improving the stress concentration on the surface defects is critical to suppress crack initiation and fatigue from starting on machine parts.

The formula for the surface factor is shown below. Depending on the surface characteristics, a higher tensile strength can create a lower surface factor. If the surface finish is very good, then the negative effect of higher tensile strength is reduced. Remember that tensile strength creates a higher endurance limit in the basic equation. So, under many circumstances, a higher tensile strength can improve fatigue strength if the surface finish is good. Let us compare two extreme surfaces, such as a forged surface versus a ground finish. For a ground finish, a 120ksi Sut material would have a Ka = .89. For a rough forged surface, the Ka would be .94. However, for a forged surface with 60Ksi Sut, the Ka would be .67 and for 120ksi material the Ka would be .34.

$$Ka = aSut \wedge b$$

Where Ka = surface factor
      Sut = ultimate tensile strength
        a = factor
        b = exponent factor

The size factor, Kb, is about the part size and varies depending on the type of cyclic force. Axial loading has no affect. Bending and torsion loading on large sizes varies from .6 to .75.

For load factor, Kc, torsion and shear is the biggest factor reduction at around .577. Axial loading is about .923. Bending is typically 1.0. Most fatigue failures I have seen in industrial settings have a bending element as the dominant cyclic force. This factor would be significant if the cyclic load is torsional, which is not as common. Don't let the load factor for bending fool you, bending cyclic stress is a major player in fatigue failures.

The temperature factor, Kd, typically will not be a reduction until the steel temperature goes above 500°F which would be a Kd .995. At 1000°F, it can be reduced to .698 so at higher temperatures the temperature factor can be significant.

The miscellaneous factor, Ke, has many elements to it such as corrosion, metal coatings (sprays, plating, etc.), fretting and stress concentrations. Some of these can be difficult to predict but the stress concentration effect is something that has been modeled. The expressions below show the relationship for fatigue adjustment factors around stress concentrations.

$$Ke = \frac{1}{Kf}$$

$$Kf = 1 - q(Kt - 1)$$

Where Kf = fatigue stress concentration factor
     q = notch sensitivity factor
     Kt = stress concentration factor

Even though fatigue is a relatively new phenomenon in engineering study, stress concentration is an even newer one. The loss of WWII warships from fatigue failure renewed the research into fatigue. It was estimated that up to 1952, some 1289 warships suffered fatigue failure, and in 1954 the first commercial jet, the Comet, broke apart in flight killing all on board. A fatigue crack initiated from the corner of the airplane window (original windows were square). The windows today are round as a result of this study of fatigue failure. Starting from that incident, Rudolph Peterson began to develop the concepts of notch sensitivity and residual stress concentrations, later published in 1974, which culminated in the Peterson stress concentration factor, Kt.[2]

The notch sensitivity factor, q, and stress concentration factor, Kf, can be found in machine design tables from various sources. These factors vary according to material, tensile strength, notch radius, shape and type of stress (bending, torsion, tension). In poor designs the miscellaneous fatigue factor, Ke, can be as low as.4 to.5 which can be a major factor in fatigue failure.

While it is good to check all these fatigue factors when doing analysis on fatigue failures, the two big reduction factors typically seen are the surface factor, Ka, and stress concentration factor, Kf, which becomes Ke.

The reality is that operational loads can be changing as well as these other endurance limit reduction factors which may not be constant over the life of the component. From this standpoint the empirical application of formulas doesn't present absolute value. However, even though predicting fatigue can be difficult, the impact of these key principles can be applied to solve fatigue failures.

Mechanical fatigue theory is something the reliability engineer can use in RCFA. There are really two key areas to focus on when it comes to solving fatigue – the cyclic stress and the material condition which will reduce the endurance limit. Once the fatigue fractography is understood through metallurgical analysis, focus the analysis on understanding the cyclic stress (magnitude, direction, source, frequency). If there is no cyclic stress, there can be no fatigue. Also focus on the part design regarding the physical dimensions and material. The combination of these two main areas make up the recipe for fatigue. Action items to correct fatigue will also be focused around these two key areas and will be highlighted in the case studies.

Most of the case studies are a high-level summary of the RCFA as there is not sufficient space to share too many details. After reading the numerous case studies in each section, a general flow of RCFA can also be seen. There will be various weights given to different sections of each RCFA case study to highlight certain aspects of the RCFA process.

3.2 CASE STUDY 1 – CHAIN FATIGUE FAILURE

FAILURE EVENTS

A recovery boiler in a paper mill was experiencing drag chain failures (Level 4) in an electrostatic precipitator. The precipitator had two separate cells with separate drag chain systems, each having four separate chain runs. Both units experienced failures in random places of chain runs. Everything in both cells is the same design and operation so the method of differences from the data did not reveal anything. The conclusion is that whatever is happening in one cell is happening in the other. In 2 years, there had been six failures (three in each cell) which had a total financial impact of over 1 million dollars.

The boiler and precipitator unit had operated for about 15 years from when installed new without any known previous failures. The chain was a rivetless design, size X-678 with a center link and two side links with pins. Failures had begun to occur in the drag conveyor on the chain center links and most failures were catastrophic-type failures. Before the final failure analysis is presented, let's summarize a few early attempts at RCFA that did not find the problem. Failure types and the component breakdown were as follows.

- Fractured Center Links – Primary Failure
- Bent & Torn Side links – Secondary Failure
- Bent Flights – Secondary Failure
- Broken Bolts – Secondary Failure

After the second failure, the primary failure mode was identified as a fractured center link. At that point, the analysis focused on that component failure. The initial failure mode of the primary failure of the center link was just identified as fatigue. Some possible causes looked at on the initial RCFA were the following:

- Chain wear – no significant issues found
- Chain misalignment – no issues found. Chain pitch tolerances were found to account for a potential 2–3" over length of chain run on new chain. Special

ordered chain was an action item to obtain closer tolerances on assembled chain length. This could potentially load one of four chain runs more than the others

- Chain maintenance – typically only failed chain pieces replaced, annual inspections completed
- Chain material issues – all chain samples had spectrochemical analysis that conformed to Microtuff chain material, material hardness was on target and showed no results off specification. One chain did show an inclusion defect at the forge line per the metallurgical analysis
- Operating temperatures – 450°F (232°C) so no fatigue reduction although ultimate strength might be reduced 10% for operating temperature
- Chain loading – no issues, motor load typically ran 65–70% all the time
- Chordal action of chain as potential cyclic force on chain. The four-tooth pitch sprocket can cause up to 18% speed changes on chain as it accelerates around the sprocket. This is worse on low tooth sprockets which were present on this one. The only problem with this theory is that the same design had been in place for the first 15 years of operation and with no failure

One conclusion of the early failure analysis completed from some samples sent to a metallurgical lab was that the failures had originated from forge line flaws on the chain link material which were severe enough on some links to initiate chain failure. One of the significant action items from the initial RCFA was to change chain suppliers if there was a material quality issue with the current one having material flaws making it vulnerable to fatigue failure. The result of using a different chain manufacturer of the same type of rivetless chain produced the same result. When the failures continued, the second round of chain RCFA was initiated. While the first RCFA efforts did not produce success, it did get the team closer to solving the failures as many things were investigated. The rest of this section is information from the second round of RCFA.

FAILURE MODE

The failure mode was high cycle unidirectional bending fatigue with low nominal stress on the chain center links ~2.5" (63.5mm) from end of chain inside. Around 75–80% of the fracture surface was fatigue and the remaining was overload at final fracture. Fatigue crack initiation started at the forge line of the chain link. One chain did have an inclusion and one didn't, but both suffered the same fate. None of the recorded center link failures were on the flight attachment links. Figure 3.1 shows the fracture surface from two different chain manufacturers with the same types of fractures.

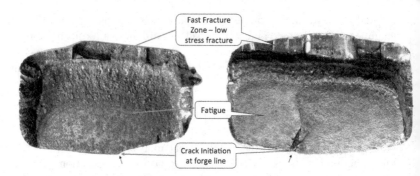

Fast Fracture
Zone – low
stress fracture

Fatigue

Crack Initiation
at forge line

FIGURE 3.1 – Chain fracture surface from two different failures and manufacturers

POSSIBLE CAUSES

There were many good things evaluated in the first RCFA. The remaining possible causes to explore a more defined failure mode after additional failures focused the second round of RCFA on two possible causes:

- Chain material defect causing crack initiation point – contributing to the cause of some failures as noted in previous metallurgical analysis but not all failures.
- Excessive chain cyclic load exceeding fatigue strength – where is the source of the force?

ANALYSIS

The chain is a rivetless chain size X-678 made from Microtuff material. The chain had a working load rating of 7,100lbs (31.6kN). The actual load on the chain was calculated at around 2,852lbs (12.7kN) at motor full load with service factor. The working load is not the ultimate strength of the chain but a safe working load limit set by the chain manufacturer to size the chain for good wear and full life. The actual drive load also never got above 70% so for normal chain operation, the load was not a factor. However, for fatigue it could depend on where the bending stress was originating.

The drive load schematic is shown in Figure 3.2. There were several drive components to connect the final chain drive with the belt drive, gearbox, and a power chain drive. The drag chain had four runs of chain.

The expected chain life is around 10 years from the manufacturer with a normal failure mode of pin wear and elongation. After the initial failures started, the chain life, cut short from fatigue failure, was only lasting about

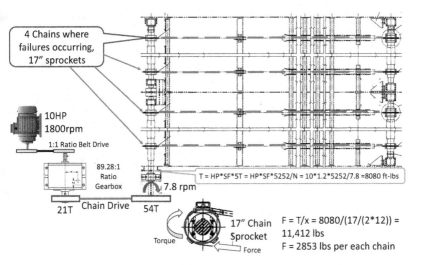

FIGURE 3.2 – Chain drive loads

9–18 months. New chain would be installed annually and between failures spot repairs were done, replacing only damaged chain.

PHYSICS OF FAILURE

The max drive load through the drive arrangement produced a torque of 8,080 ft-lbs (10,955N-m). The physics and force around the sprocket would produce 2,853 lbs (12.7kN) per chain. So, the chain has this force pulling on the center link.

The source of the bending cyclic stress on the center link was found to be the worn sprocket, as shown in Figure 3.3. Doing the force analysis on the chain bending moment from the sprocket where the actual bending occurs (~2.5" from end), the moment from drive is ~4,032 in-lbs (455 N-m). The resulting bending stress on the center link would be ~47,000 psi (324 MPa). The endurance limit for this chain is estimated to be ~21,000 psi (145 MPa), which is less than the cyclic stress, so a finite life is expected. The estimated number of load cycles was in the 9–15 months' range under these load conditions, which is near where many of the failures occurred (4–6,000,000 cycles). The S-N curve would look something like Figure 3.4.

None of the center links with attachments ever failed. The attachment links fill in the gap between the chain sections (top and bottom) essentially making the center link like a solid member. The attachments stiffen and support the center

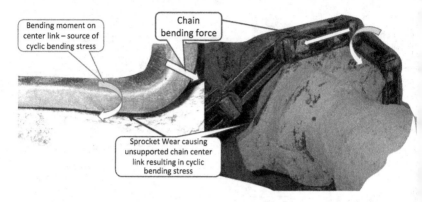

FIGURE 3.3 – Sprocket wear – source of bending stress on center link

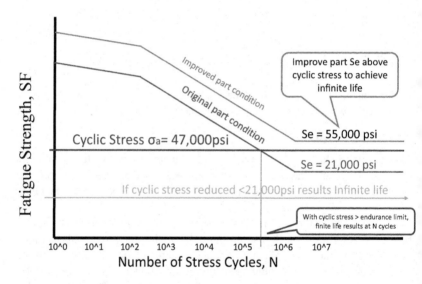

FIGURE 3.4 – S-N curve for chain with alternating stress

link which virtually increases the endurance limit of the center link to infinite life. This was another clue in evaluating the source of the bending cyclic stress.

ROOT CAUSES

The physical root cause of failure was worn chain sprockets, which led to inducing a cyclic bending stress to the chain center link. This bending stress exceeded the fatigue strength or endurance limit of the chain link, resulting in bending fatigue failure. The latent root cause could be identified as insufficient inspections to identify the sprocket wear. The chain system was inspected annually by contractors who perform that specific work, but no one was aware of the sensitivity of the specific design regarding fatigue failure.

ACTION ITEMS

Below are some of the possible action items to prevent root causes.

- Replace all sprockets in the precipitator. New sprockets also had some design changes such as making the sprockets a little harder to decrease wear and making the sprockets split for ease of installation.
- Special order matched chain sets with tighter length tolerances. This moved full length chain standard tolerance from +/-2.75" to +/-.5". This may have been overkill but for the first fix replacement this was an insurance move for greater precision of chain operation and loading.
- Surface harden the chain to improve fatigue resistance. A shot peening process was used to improve the surface factor and endurance limit on the new forged chain.
- Revise existing chain inspection annual PM to include sprocket inspection for wear. Share chain sprocket wear and chain RCFA with the contractor doing the inspections.
- Non-destructive test any previously reused chain links for cracks on annual outages. Did not execute this action item as all chain was replaced when sprockets were replaced.
- Redesign the chain system to a different style chain to move away from the rivetless forged chain to one with improved fatigue characteristics. Did not execute this item since root cause was found. This also saved a $300K capital project for a chain upgrade.

SUMMARY

The conveyor had operated for around 15 years before any chain failures showed up. It took that long for the chain sprocket wear to be significant enough to cause the bending fatigue and a finite chain life of 1–2 years. There were some chain links that didn't fail, likely because they did not have as poor a surface condition at the forge line on some links. Some failed links had material flaws which further reduced the fatigue strength. Some links failed even though they didn't have material flaws. The forge line is present on all forged chain links as part of the manufacturing process. The item that changed was the cyclic load on the chain center link as the sprocket began to wear. This is a good example of how operating mechanics can change over time when conditions change.

Multiple failures with metallurgical analysis revealed lots of good information but neither could identify the root cause of failure but only that the crack initiated at the forge line. The scope of metallurgical analysis was mainly to understand the fracture surface fractography and to identify the material grade and quality. The failure took further analysis using key elements of fatigue failure to solve and get to the root cause.

3.3 CASE STUDY 2 – CONVEYOR SHAFT FAILURE

This case study has two different failures, with two different failure modes, which will both be briefly discussed.

FAILURE EVENT 1

A bark reclaimer system has an overhung gearbox drive which experienced shaft failures (Level 4) about a year after a shaft and gearbox were changed.

FAILURE MODE 1

Rotating bending fatigue with low stress and stress concentration as fracture surface is shown in Figure 3.5. The failure occurred about a year after the new shaft was installed. No other significant damage was experienced in this failure.

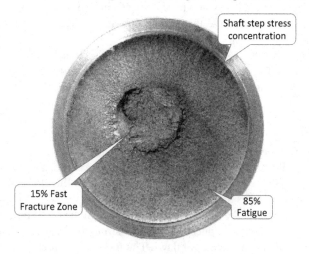

Shaft step stress concentration

15% Fast Fracture Zone

85% Fatigue

FIGURE 3.5   – Failure mode 1 – shaft fracture surface

FAILURE MODE 2

Rotating bending fatigue with low stress concentration occurred, as shown in Figure 3.6. The fatigue fracture crack initiation was at the end of the keyway corner.

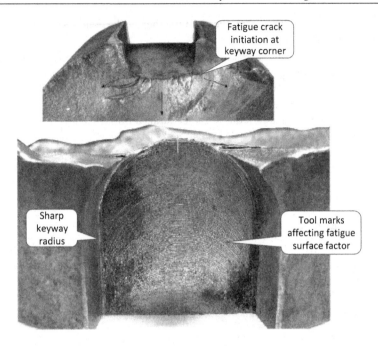

FIGURE 3.6   – Failure mode 2 – keyway shaft fracture

POSSIBLE CAUSES

- Excessive, cyclic bending stress on the shaft – bent shaft, insufficient shaft design (shaft diameter, material, etc.)
- Shaft defect reducing the fatigue strength (endurance limit) of the shaft.

ANALYSIS

The conveyor mechanical drive including motor, coupling and baseplate is cantilevered off the reclaimer conveyor shaft. The total weight of the drive assembly is ~6,500lbs (28.9 kN). The drive consists of a 75HP (56kW), 1800 rpm motor and gearbox with a 97.13:1 ratio and a 1.53 Service Factor (SF). The stress cycles on the shaft up until failure were around 10,000,000 (10^6) cycles, which is around 375 days at design speed. The original shaft was a 1045 carbon steel shafting (90ksi tensile).

Looking at the fracture surface, the fast fracture zone (FFZ) was only about 10% of the total shaft volume so normal bending stress was low. The source of the bending stress is the cantilevered load of the drive so, without redesigning the entire system, there is little that can be done about the bending load on the shaft.

In the original shaft design, the drive system had operated for many years without any failures. The shaft endurance limit was higher than the cyclic stress which gave the shaft infinite life. Using the POF principles, the fatigue factors that reduced the endurance limit were the surface factor and stress concentration factor. These factors reduced the endurance limit by ~50% which gave a finite fatigue life on the shaft. The shaft had surface roughness of ~150–200Ra. The shaft radius at the shaft step was <1/64" (.41mm).

When some routine maintenance was performed, replacing the shaft, it was not an original manufacturer-supplied shaft, but one that was built from a local shop without any detailed building drawings supplied.

ROOT CAUSES

The physical root cause was the shaft stress concentration from the insufficient shaft radius and machine tool marks at the shaft step. The second failure was a result of an incorrect keyway radius and surface finish (tool marks). The latent root cause was shaft design and machining when the shaft and gearbox were replaced.

ACTION ITEMS

- Manufacture a new shaft with a redesigned shaft radius.
- Make a shaft from a 4140 turned ground polished round stock with a tensile strength of 120ksi. Polish shaft radius to a polished finish (<16Ra). Figure 3.7 shows before and after photos of the shaft step.

FIGURE 3.7 – Original shaft and redesigned shaft

- Redesign the shaft keyway to have a large end radius instead of a sharp radius.
- Modify the shaft drawing to reflect changes and communicate with new shop drawing.

SUMMARY

Once these conditions were corrected on the shaft design, the reclaimer system once again operated for many years without any failures. This is another reminder to see any design changes or maintenance work prove itself out for a period of time past infant mortality. It also shows that design details matter in long-term reliability.

NOTES

1 Shigley & Mischke, Mechanical Engineering Design, McGraw Hill Publishing, 5th ed, 1989, p.275–290.
2 Fred Eberle, The History of Bending Fatigue, Part 1, Gear Solutions Magazine, April 2015, p. 18–19.

Key Factors in Bearing Failure

4

4.1 FUNDAMENTALS OF BEARING FAILURE

This section will focus only on antifriction roller bearings which include ball bearings, spherical bearings, cylindrical bearings, and tapered roller bearings. Bearing life can be a complicated subject with many variables. There are many failure modes for bearings such as fatigue, brinelling, false brinelling, corrosion, fluting, pitting, smearing, peeling, fracture, particle denting, wear, fretting, and several others. Each failure mode has its own possible causes and root cause of failure.

A common reference of L10 bearing life is expressed in the equation below, which can identify the physics of failure from these variables. L10 fatigue life means that in fatigue tests, 90% of the bearings achieved the expected life under test conditions. L1 life would mean 99% of bearings achieved the expected life. L10 is the industry standard most often referenced. This bearing life estimate is only valid for general fatigue failure modes. Other failure modes really don't have a way to predict life other than with repeated field studies.

$$L10 = a1a2a3\left(\frac{16,667}{n}\right)\left(\frac{C}{P}\right)^{p}$$

Where L10 = fatigue life of bearing

        a1 = reliability factor, a1 = 1 for L10

        a2 = material factor, typically a2 = 1

        a3 = lubrication factor

        n = rotational speed, rpm

        C = basic dynamic load rating, lbs

DOI: 10.1201/9781003248675-4

P = equivalent dynamic load, lbs
p = factor for bearing type, ball p = 3, other roller bearings
   p = 10/3

The big three bearing life variables are load, lube, and installation. We will briefly discuss these three variables and then support some examples in case studies.

Bearing Load

One of the functions of bearings is to support the operating load. Loads can be in multiple directions such as radial or thrust (axial). Radial loads can be in multiple planes and axial loads can be in either direction. Bearing loads can also change during startup or the operating conditions of the equipment they are operating. The equivalent bearing load in the L10 equation is a combination load from all the loads acting on the bearing. Most bearings can handle different types of loads and some can only handle single directional loads. Cylindrical bearings only handle radial loads. Most bearings have some minimum bearing load which is typically in the 2–3% of dynamic load rating for the bearing. Excessive loads, extreme small loads, and wrong types of loads can all cause bearing failure.

Lubrication

Lubrication has been reported to account for more than 50% of bearing failures. Lubrication is something that factors into every moving element. Lubrication is one of the first elements to be investigated for a bearing failure. Lubrication can be tricky, and many root causes of failure even not related to lubrication will reveal lubrication issues on bearing failures. Loading or installation root causes may show up as a lube deficiency on the bearing.

There are many things to get right in lubrication, such as right product, right amount, right location, right delivery, and right quality. Miss any of these and failure may begin. The most important lubrication parameter to get right is lube viscosity. It is usually not difficult to find the target lube viscosity for most applications. Many equipment nameplates such as motors, gearboxes and pumps may have lube specifications listing the lube viscosity. The type of lube used (oil or grease) is also important. This can depend on the equipment design or the bearing specification. Some bearing applications may be able to use either. The additive package is another important piece. It can either be a rust and oxidation (R&O) package or anti-wear (AW) or extreme pressure (EP). Different applications and conditions will drive different additive packages.

Installation

Many failure modes and root causes of failure can be attributed to installation issues. Some have said bearings do not die, they are murdered. We will see some murder investigations from a few case studies here in this chapter. Bearings are precision components and must be installed with the greatest care.

Bearing mounting can have many methods, not all of which are universal in every application. Cold mounting bearings with special tools are only possible on small bore bearings (typical 70mm max for cylindrical bore and 240mm max for taper bore). Cold mounting for a cylindrical bore requires special sleeves to fit the size of bearing installed. Cold mounting for a taper bore requires the proper spanner wrench for the bearing nut. Temperature mounting is typical for cylindrical bore bearings and can include an induction heater or oil bath heater. Care must be taken to not overheat the bearings or leave residual magnetism during this type of installation process. Hydraulic mounting includes using a hydraulic bearing nut to drive a tapered bore bearing onto a shaft.

Bearing mounting must achieve the proper fit to the mounting shaft and remove the radial internal clearance (RIC) to achieve correct installation for full bearing life. The RIC is the clearance between the rolling elements and the inner or outer race. For cylindrical bore bearings, this is determined by the fit of the inner race to the shaft for inner race rotation applications. This fit will be some magnitude of interference fit. The outer race fit will then typically be some magnitude clearance fit. Measuring the shaft, bearing, and housing dimensions are key in assuring that the bearing fit is within target.

For a tapered bore bearing, the final mounting RIC is determined by how much the bearing is pushed up on the tapered journal. Using the correct manufacturer bearing clearance card is critical for each specific installation as there is not a standard RIC reduction. Undermounting or overmounting a bearing can have extreme effects on the life of the bearing.

There are also many methods for installing these tapered bore bearings. Some bearings or applications must only use specific methods. For example, a self-aligning ball bearing must use the angular drive-up method. There is no way to measure RIC on a ball bearing. Below are some of the most common methods for installing taper bore bearings.

1. Feeler gauge method – Using the unmounted RIC as a starting point, measure the bearing RIC as the bearing is driven up the shaft until the desired RIC reduction is achieved. For example, on a 1:12 taper the axial drive up distance =16 times the RIC reduction.
2. Axial drive-up method – From a neutral starting position on the shaft, using a dial indicator on the bearing inner race, drive up the bearings to the desired axial distance to achieve the target RIC reduction.

3. Angular drive-up method – From a neutral starting position on the shaft, drive up the bearing to a prescribed angular clock degree for that size bearing.
4. Hydraulic drive-up method – The bearing is driven up the shaft, using a special hydraulic nut, to a prescribed pressure (to establish a neutral starting position) and then a prescribed axial distance using a dial indicator to measure the final travel.

There are other special bearings, like printer bearings, which require an even more different procedure to measure the bearing rolling friction. These bearings are preloaded bearings and can't use the methods above for correct mounting. All these methods require specific tables for a specific bearing to complete the precision installation.

Some typical bearing failure modes are listed below.

1. Fatigue caused by cyclic rolling stresses on the bearing elements. Fatigue can be surface fatigue or sub-surface fatigue. Surface fatigue is where flaking and cracking originate at the rolling surface. The sub-surface originates below the surface with micro cracks.
2. Wear is where the surface asperities of the two rolling or sliding surfaces contact and remove material. The wear can be abrasive wear or adhesive wear. Adhesive wear is where material is micro welded and tears away, transferring from one surface to another. Contamination or a poor lube film may cause abrasive wear.
3. Corrosion is a chemical reaction on the metal surface. Corrosion can be from moisture where the surface is oxidized. Corrosion can also come from fretting where micro movements and friction cause corrosion that is typically seen on bore and outer races.
4. False brinelling involves cyclic vibration and micromovements in the bearing and can be like fretting corrosion but shows up on bearing raceways.
5. Electrical erosion is where electrical current removes material from the raceways. It can be electrical pitting- or electrical fluting-type damage.
6. Plastic deformation is a permanent deformation exceeding the yield strength of the material. It can come from overload, called true brinelling, or from debris denting.
7. Fracture occurs when the ultimate tensile strength of the material is exceeded. It can be an overload fracture, a fatigue fracture or thermal cracking (heat cracks).

There are volumes of books on each element of bearing reliability, but load, lube and installation are three of the key areas to investigate once the failure mode of bearing failure has been carefully identified. The following case studies will highlight some of these key elements.

4.2 CASE STUDY 1 – COMBUSTION FAN BEARING FAILURE

FAILURE EVENTS

For many years there had been infant mortality Level 3 and 4 types of chronic failures on the bearing assembly for two combustion fans on a paper machine. Some units ran for 6–12 months while some only lasted 1–2 months. There are two fans with one spare bearing housing assembly for both. Failures occurred on both the wet end and dry end fans. The fan systems are identical. It was the last major failure where extra measures were taken to investigate the failures thoroughly to find the root causes. While only one failure mode was investigated, all other potential issues with the fan were also evaluated. This bearing assembly only ran for less than a month. Vibration analysis typically called

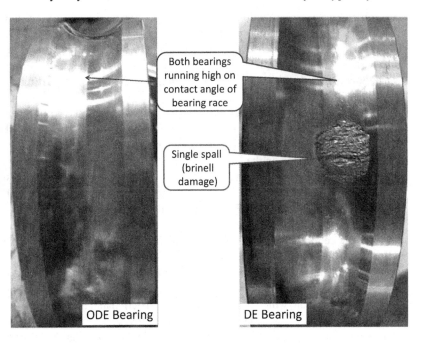

Both bearings running high on contact angle of bearing race

Single spall (brinell damage)

ODE Bearing   DE Bearing

FIGURE 4.1   – Fan bearings

out bearing issues, so maintenance was accomplished on scheduled outages. Repair cost was about $25,000 annually.

FAILURE MODE

• High bearing contact angle (on outside of each bearing) and 360° load zone on outer race of both bearings.
• Single spall on inner race in roller path on floating end (fan end) of bearing assembly, as shown in Figure 4.1.
• Severe scoring damage on shaft and bearing bore surface.

POSSIBLE CAUSES

• Insufficient shaft fits
• Insufficient lubrication
• Assembly errors
• Shipping and handling damage
• Incorrect parts – shaft and ball bearings 6220 C3, bearing end covers

ANALYSIS

The combustion fan is a radial blade fan with 42" wheel with a direct drive 350HP (261kW), 3600rpm motor. The fan produces air flow of 10,000CFM. The baseplate is equipped with spring isolators for vibration dampening. The fan bearings are both 6220C3 deep groove ball bearings. The ball bearing has a dynamic load rating of 27,900lbs (124kN) and a static load rating of 20,900lbs (93kN).

The shaft fits are within targets of.0001/.0015"T (.00254/.0381mm) and did not appear to be an issue with this failure mode. The lubrication called for an ISO 68 R&O oil which is what was being run in the bearings; however, there was an issue noted on the amount of oil used. Maintenance had been filling the unit to near the top of the bullseye sight glass. The reason given was that that was the only way to get oil to the center of the bottom-most roller of the bearing which in most cases is the correct oil level. A red flag on any bullseye oil sight glass is when the oil level is not in the center. The target oil level is always the center of the bullseye. The correct oil level was the center of the bullseye, as shown in Figure 4.2. The manual specified 6 quarts of oil. This oil level is below the bearing roller center but, for high-speed application, an oil flinger design was used to supply oil to the bearings. The bearing was running in a flooded condition, which would create higher friction and heat generation.

There were a few assembly errors discovered that added to the operating conditions of the fan operation. The bearing assembly had different bearing housings for the floating and fixed ends. It did not appear to impact bearing spacing or location. The only affect was the position of the oil flinger being inside or outside the housing. The other assembly issue was that the wavy spring was left out of the floating end (the fan end).

The single spall damage is one that appears to be a Brinell-type damage from impacting. Initially it was not known the source of the impact force. One thing to note from the failure mode of the spall mark is that the location is at a high contact angle on the floating bearing. A static condition and even an operating condition should have the floating bearing loading the bearing more in the center of the bearing. This was another red flag clue in the investigation.

Incorrect parts were looked at for the bearing assembly. The original design calls for a C3 bearing which was the correct bearing used. Talking with maintenance technicians who installed the assembly that failed, they noted that it was harder to turn than they felt it should have been when installed. After failure, the bearing arrangement was disassembled. The wavy spring was found to be missing on the floating end because it would not fit when assembled, which was another clue.

Further investigation revealed that the floating end bearing cover was incorrect. The manufacturer made both bearing end covers the same, but gave special instructions for the end user to machine off the floating end bearing cover before assembly, as seen in Figure 4.2. During assembly, the machining operation to reduce the bearing end cover lip was not completed, which left the bearing assembly with two held bearings.

Looking at the POF, the formula for thermal growth is shown below. The thermal growth on this shaft assembly was calculated ~.007" (.178mm) which yielded a significant bearing load on the bearings, causing the high contact angle. The high contact angle also meant that if there was impact damage it most likely occurred after the bearing assembly. The result was that the assembly was preloaded.

$$\Delta = \propto L\Delta T$$

Where Δ = Thermal growth, change in length, inches
$\propto$ = Thermal growth coefficient, 1/°F
L = length of member, inches
ΔT = Change in temperature, °F

One other observation is the bearing assembly that had the incorrect floating end cover had a few runs where the assembly ran for around a year or so before it was changed. The ABEC/ISO 5-ring width tolerance for this bearing is 0.000/-.025" (0.0/.635mm) so the ring width on a different bearing could be in specification and be as much as .025" (.635mm) less width. That means that the bearing assembly could have handled the thermal growth and still had some float without heavily axial loading the bearings. There was no way to prove this as past failures were not documented.

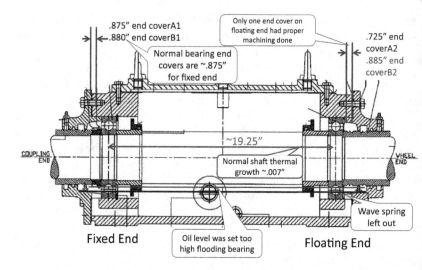

FIGURE 4.2   – Bearing housing assembly

ROOT CAUSES

The physical root cause was that the bearings were preloaded at assembly due to incorrect floating end bearing due to an incorrect end cover. With no floating end, thermal growth resulted in excessive bearing loads and possible Brinell damage. The bearing housings being swapped in the monoblock also positioned the fan slightly differently in the housing, which created a greater thrust load on the bearings. The wave spring was left out of the assembly because it would not fit. The latent root cause was that there was no detailed rebuild procedures for the shop to assemble the bearing housing assembly correctly.

POSSIBLE ACTION ITEMS

- Investigate the warranty on the bearing assembly on the next repair.
- Machine one of the bearing end cover plates from ~.875" (22.23mm) to ~.750" (19.05mm). Reassemble using a wave spring on the floating end.
- Set up a rotating assembly as a storeroom spare and correct the bill of materials. This was previously being rebuilt by the area manually instead of through the storeroom process.
- Add an oil level sight glass at one drain port to mark the correct oil level as a secondary oil level indication.
- Establish a rebuild procedure for the fan rotating assembly (bearing fits, bearing end caps) and communicate to the rebuild shop.

SUMMARY

The failures with the one assembly increased when looking back at the maintenance history when new bearing end covers were installed several years earlier. During the rebuild, it was not known that, even though the bearing end covers were the same, the floating end had to be machined to achieve the necessary clearance. Once this was corrected, the fan bearing temperature dropped by nearly 40°F (4.4°C). Overall reliability improved on all bearing housing units as well once the oil levels were adjusted to the correct level.

4.3 CASE STUDY 2 – SUCTION ROLL BEARING FRACTURE

FAILURE EVENT

During normal operation, the paper machine experienced a sheet break. During thread-up, operators found the suction press roll locked up and not turning. The machine was shut down and the roll was changed. This was a one-time event and caused significant machine downtime with the total cost of the failure over $200,000.

FAILURE MODE

The bearing failure was a catastrophic failure as nearly every part of the bearing had sustained damage and became a 1000-piece puzzle upon disassembly. Initial observations of the bearing and roll during disassembly made the hope of finding a root cause look slim. From roll shop records of the previous 35 years and reports from numerous roll shop veterans, there had never been a failure like this. A few elements of the failure mode are listed below.

- There were multiple fractures in the outer and inner race of bearing, more damage on the inside row of the bearing.
- The cage fingers were fractured from overload and not worn on the inside row.
- The inner race and outer race spun on the journal and housing.
- The outer race fracture was circumferential in some areas along W33 groove.
- There were fractured rollers on the inside row.

POSSIBLE CAUSES

- Lubrication deficiency – wrong lube, wrong amount.
- Bearing load issue.

- Bearing issue – defect, manufacturing flaw, metallurgical issue, installation issue.

ANALYSIS

This roll was a suction press roll that was shared between two similar machines. The original shell for the roll was a 1979 bronze shell. That shell suffered a corrosion fatigue fracture 9 years earlier. A new duplex stainless steel shell was purchased to replace the bronze shell. The roll had just been rebuilt the previous year, ready for installation in the machine. The roll ran for about 9 months in the machine prior to failure.

The roll is a 51" (1295mm) diameter roll and runs around 3000fpm (225rpm) in the machine. The bearing is a 230/600C3 spherical bearing. The bearing application is an outer ring rotation bearing design. Bearing interference fit is on the outer race and a clearance fit is on the journal. Due to past issues with bearing creeping on the journal and causing damage, the design for all these types of rolls was changed about 10 years earlier to have a round keyed inner race.

There had been no changes in lubrication for probably 10–15 years so it was unlikely that this was a major factor in the failure. A synthetic grease with 220cst viscosity was used, which is the recommended lubricant. The bearing was being greased manually every 4 weeks with 80oz (2366cc). For bearings greater than 300mm bore, the bearing manufacturer recommends continuous greasing (this bearing is 600mm bore). While grease frequency is not ideal, this has been done for many years without catastrophic failure. The bearing was greased in the previous 24 hours from the failure and may have been a small contributor to the failure with extra temperature and stress on the bearing. The roller path and cage on the outboard side of the bearing showed no signs of lube stress; therefore, insufficient lubrication was not the root cause.

The bearing has a dynamic load rating of 1,243,000 lbf (5529 kN) and a minimum load of 37,000 lbf (165 kN). With the 450 PLI (pound per linear inch) nip and 23" (584mm) Hg of vacuum, the estimated resultant bearing load was around 82,000 lbf (365 kN) so the bearing was above the minimum load and considered a lightly loaded bearing.

The bearing metallurgy showed no significant issues. The bearing races were in the 56–58Rc range and the rollers were 62Rc. Other metallurgical analysis of the fracture surface did not reveal any significant issues. It did conclude that the darkened area of the fracture surface which was on the outer race outer diameter (OD) and inner race OD were fatigue cracks that were darkened by the baked-on hydrocarbons of the lubricant. The fast fracture zone did not have that type of fracture surface.

Vibration data was taken 10 days before failure and no defects were noted in the bearing and conditions were normal. If there had been a significant bearing defect in the bearing or assembly, it would have shown something to that effect in the early months of operation.

The roll had a new journal on this bearing fit. From shop records the bearing had a .004" L (.102mm) bearing fit on the shaft as was the target. The bearing outer race fit was .007" T (.178mm) which was at the maximum for this roll design. The shell fit will reduce this housing fit by some amount, making the bearing fit tighter. The bearing manufacturer recommends taking the housing fit dimensions with the head mounted in the shell to gain the cumulative effect of fit, but most shops do not perform this extra step. Since the change to the stainless shell from the original bronze shell, this head to shell stress at operating temperature has not decreased as much. The bronze shell thermal expansion reduces the fit more than the stainless shell which is the reason for the difference in target fits.

The inner race fractured from the outside in the keyway. The keyway was not an initiation point of the fracture. There are several places that would have been stress concentration points for crack initiation. The only place where it did appear that crack initiation likely occurred was near the W33 lube groove on the outer race. Another note was that this fracture on the inner race occurred right over the key, which is out of the load zone for this roll which means it would have clearance between it and the journal.

The circumferential crack on the outer race indicates a differential stress (load profile) on the outer ring from the inside to the outside of the bearing. This outer ring fracture did occur mostly in the weakest section of the outer ring, which is the center ring W33 oil groove and was circumferential.

There was more fretting on the outside row of the outer race and there was evidence that the roller path on the inside row was more heavily loaded than the outside row. All the evidence was starting to point toward an issue with the bearing housing taper or head fit.

After verifying the head fits from the shop rebuild sheets, the root cause was found, as shown in Figure 4.3. The head size at station I (the outside row of the bearing) was .004" (.102mm) smaller than the shell with a .012" T (.305mm) fit. The head size at station II (the inside row of the bearing) was .008" larger than the shell with a .024" T (.610mm) fit. With a .007" T (.178mm) fit already and .024" T fit added to station II, the bearing was severely overstressed. The maximum RIC to be removed on the bearing was .011/.013" (.279/.330mm) which station II greatly exceeded. Another contributor was that the target fit used was .002" (.051mm) tighter, which was correct for a bronze shell instead of the stainless shell being used.

ROOT CAUSE

The physical root cause was that the excessive head/shell interference fit (taper) at Station II caused an overstress condition on the inside row of the bearing's inner and outer races. The inside head/shell fit crushed the bearing inside row. The shell material changes also contributed to the bearing load.

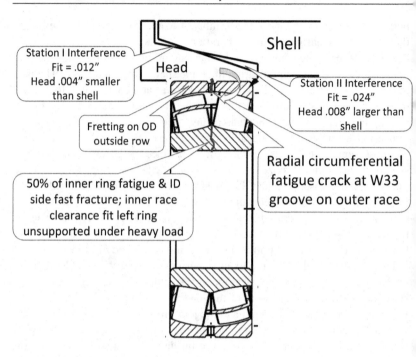

FIGURE 4.3 – Bearing fracture

POSSIBLE ACTION ITEMS

- Machine and correct the head and shell fits on this roll.
- Modify the roll rebuild sheets to improve the head fit procedure to reduce human error.
- Review findings with all roll shop technicians who rebuild suction rolls.
- Modify the greasing of the bearing so that greasing occurs every 2 weeks, thus spreading out the grease amount hitting the bearing.
- Investigate the continuous greasing system for the press section.

SUMMARY

This was the first stainless shell for the machine rolls, so the roll shop was not aware of the lower targets on a stainless steel head. Likely extra shims on previous rebuilds kept the assembly head stress low enough to not fracture the bearings. Lubrication added stress to the bearing. The circumferential cracks, as well as being out of the load zone, hid many of the indicators on vibration analysis, not to mention an indirect path to the bearing since the bearing is mounted inside the head in the rotating shell. The rotating outer race was also different than most bearing installations. When the right conditions align, failure can occur.

4.4 CASE STUDY 3 – PULPER MOTOR BEARING FAILURE

FAILURE EVENT

A batch pulper motor had a planned outage job to change a bearing that had an early-stage defect identified from vibration analysis. At the completion of the job on start-up, the motor showed signs of failing instantly. The motor in minutes went to ground at the final failure. This initial motor failure shut down the 4160V feed and shut down the entire de-ink plant. The cost of the failure was ~$150,000.

FAILURE MODE

The failure was a Level 4 functional failure. This was not a chronic failure as it was the first maintenance performed on this new motor.

- Primary failure mode: The drive end (DE) bearing spun on the shaft and housing. Wear continued until the motor rotor shorted out with the stator. The bearing was essentially destroyed as the motor ran all the way to catastrophic failure.
- Secondary failure mode: The bearing end covers were worn due to shaft rub. The bearing cage was torn apart. Motor rotor laminations were pulled due to contact with the stator. Snap ring bent and was damaged.

POSSIBLE CAUSES

- Bearing contamination
- New bearing defect
- Lubrication deficiency
- Mechanical binding of bearing assembly
- Bearing fit issue

ANALYSIS

The motor was a 900HP (671kW), 1200rpm TEFC, 4000V motor. The motor DE bearing is a 6230 deep groove ball bearing lubricated with PolyrexEM grease. The original motor was an 800HP (597kW) open frame motor. The pulper drive motor was upgraded about 4 years earlier to the 900HP. This bearing change was the first maintenance that had been completed on this upgraded motor.

The bearing change was due to an early-stage bearing defect identified from vibration analysis. Due to the location of the motor and cost, it was decided to change the motor bearing in place in the field. During the bearing change there was a washup event that sprayed water down onto the motor floor. The motor and bearing were protected during this short event and did not cause an issue but it is worth noting, due to the infant mortality nature of the failure. Bearing contamination was not believed to be a factor in the failure.

Since this is a deep groove ball bearing there is not an easy way to inspect the new bearing other than a good visual and feel of the bearing. Nothing was noted from technician who installed the bearing initially.

The new bearing installed was hand-packed with the correct new grease specified for the motor. The technician took a picture of the bearing after hand-packing with grease before installing the bearing end cover. Lubrication deficiency was not the root cause of failure.

While there was mechanical wear on the bearing end covers, the wear appeared to be in line with the shaft dropping down as the bearing failed. This was secondary damage on the bearing end covers. If there had been binding on the end covers, which could have generated enough heat so that the grease ran out or degraded enough, then it could have been a start point for catastrophic failure.

The bearing fit at the time of the bearing change was not known due to the technician not taking a micrometer measurement of the motor shaft and bearing. Finding a definitive root cause looks unlikely but the investigation was still not complete. Since this is a deep groove ball bearing there is no way to verify RIC. Further investigation had feedback from the technician that the bearing had to be heated up twice and even then, the bearing would not slide on the shaft without extra force, which was a red flag.

The target bearing fit for this bearing was an m5 fit which is .0006/.0023" T (.0152/.0584mm). If the bearing heater was heated up to 220°F, then the bearing bore would have grown by .004" (.102mm). That means that there would have been .0017" (.0432mm) clearance for the bearing to easily slide on the shaft which was not the case. The shaft diameter targets were 5.5131/5.5124" (140.03/140.015mm) to produce the fit range for this bearing. The bearing installation seemed as if the shaft was oversized, but how could this be since it was a new motor, which also ran ~4 years without failure?

If the shaft was oversized.001" (.0254mm) and the new bearing was at the minimum bore, the resulting shaft fit could have been.0036" T (.0914mm). This could have potentially removed all the bearing RIC, especially if the bearing was in the lower range of RIC (normal range is.0007/.0019"). This would cause high friction which could lead to catastrophic bearing failure, as was experienced. The reason the initial bearing might not fail due to an oversized shaft is that the bearing bore was likely near maximum bore which would have left the bearing fit in the acceptable range, as shown in Figure 4.4.

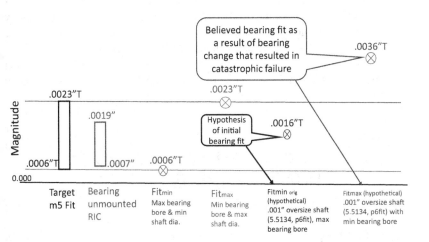

FIGURE 4.4   – Bearing fit scenarios

ROOT CAUSE

Excessive bearing fit removing RIC was found, which locked up the bearing and it spun on the shaft. The latent root cause is the original design/manufacture of the motor shaft and an insufficient procedure for motor bearing change by not measuring shaft fits.

POSSIBLE ACTION ITEMS

- Repair damaged motor to be a good spare motor.
- Investigate bearing and motor temperature trending for new motors. Some of this was not hooked up on existing motors.
- Investigate Multilin alarm communication to the distributed control system (DCS). When temperature alarms are added, communication with the DCS

is necessary otherwise operators will never see the alarm in the control room.

- Improve job preparation to eliminate contamination during the job.
- Establish bearing mounting fit information as part of the bearing change procedure.

SUMMARY

While a definitive root cause may not have been found, there was much circumstantial evidence as to the potential root cause. There were also many good action items going forward to improve the potential for avoiding future types of failures from potential root causes. This example was a good case to show how important bearing measurements can be, especially considering manufacturing tolerances when dealing with small measurements making the difference between operating reliably and infant mortality failure.

4.5 CASE STUDY 4 – VACUUM PUMP BEARING FAILURE

FAILURE EVENT

A large liquid ring vacuum pump had chronic drive end bearing failures. Three failures had occurred within about 9 months from when a rebuilt vacuum pump was installed. The failure qualified as infant mortality failure. The failures were Level 2 failures, as vibration analysis identified the failures and bearings were changed while on a scheduled machine outage. The failure impact was labor and materials which cost ~$30,000. Without a good vibration analysis program these failures could have been 10 times the cost impact.

FAILURE MODE

Spalling of the outer race (driven side row) of the double row tapered roller bearing on the drive end of the vacuum pump. Load zone was very large on the failed side of the bearing > 180°. The first failure also had a cracked outer ring, as shown in Figure 4.5.

POSSIBLE CAUSES

- Lubrication deficiency – insufficient lube, contamination
- Bearing issue – metallurgy, manufacturing defect, housing bore issue

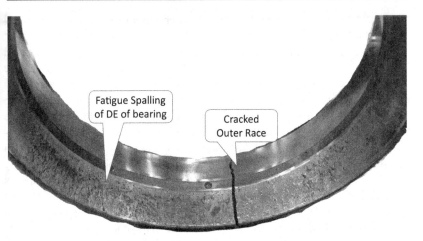

FIGURE 4.5 – 1st failure outer race

- Misalignment – bearing housings or pump and driver
- Thrust loading – vacuum pump or gearbox issue

ANALYSIS

The vacuum pump with the bearing failure is the first pump in a drive train with two pumps. The vacuum pump is driven by a 2500HP motor through a single reduction parallel shaft reducer. The vacuum pump drive end bearing is the held bearing in the drive train. Both vacuum pump bearings are double row tapered roller bearings.

Lubrication is done by local continuous grease units. The bearings have all been found to have plenty of good quality grease and in good condition. The vacuum pumps have not been leaking and flooding the bearings with water so contamination is not a significant factor. The correct lubricant is being used.

The vacuum pump experienced three bearing failures within 12 months. It is unlikely that the same bearing defect was present on all three bearings due to a manufacturing defect. Bearings were very clean on all condition monitoring right after installation. Vibration analysis revealed a bearing in good condition without any defects initially, but these showed up after a few months.

Since these are tapered roller bearings, alignment is very critical. Tapered roller bearings can't handle any misalignment. The pump was installed and laser-aligned. It was not found with any significant misalignment at the time of the failure. With the relationship between the two bearing housings, it was not known if they were line bore true to each other. After the second failure, a new drive end housing was installed but a third failure still occurred.

Once the third failure was identified, the shaft/coupling gap was inspected closely at the next bearing change. The finding was that the shafts were very close and were impacting each other. Inspection of the vacuum pump shaft revealed an impression of the gearbox shaft and keyway on the vacuum pump shaft, as shown in Figure 4.6. This was the source of the thrust load on the vacuum pump bearing. Further inspection revealed that the target shaft/coupling gap should be .250" but the as-found coupling gap was probably less than .020". Potential thermal growth of the two shafts was around .012".

It appeared that the shaft length changed when the new vacuum pump was installed. The shaft tolerance on this vacuum pump is +/-.125" (3.175mm) so a total .250" (6.35mm) potential from previous vacuum pump if both were at opposite ends of the tolerance. Not knowing the shaft coupling gap after the second failure, it was questioned why the gearbox bearings were not failing since it would see the same thrust load as the vacuum pump bearing. Further evaluation of the gear thrust revealed that the thrust for the gearbox was toward the vacuum pump. As a result, this would have unloaded the gearbox bearing and loaded up the vacuum pump bearing once the two shafts made contact. The gearbox thrust load and bearing end play would be the source of the thrust loading.

ROOT CAUSE

The physical root cause is an insufficient shaft/coupling gap between vacuum pump and gearbox, causing a thrust load on the drive end vacuum

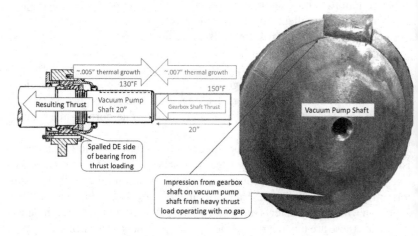

FIGURE 4.6 – Shaft thrust mechanism

pump bearing. The latent root cause is installation error from an incomplete procedure calling out the shaft gap. A contributing factor was the vacuum pump shaft design tolerance and field installation where the vacuum pump was bolt bound.

POSSIBLE ACTION ITEMS

- Replace vacuum pump with shorter shaft.
- Modify vacuum pump feet for oversized holes with eccentric sleeve to solve the bolt bound issue.
- Edit the vacuum pump installation procedure for shaft/coupling gaps.

Summary

Infant mortality failures bring attention to what changed, such as installation issues. In this failure, an entire vacuum pump was changed but the key clue was the failure mode detail in that the bearing experienced a heavy thrust load on one side of the bearing every time and it was in the opposite direction to the normal thrust load. Once the shaft gap was corrected no more failures occurred. When technicians installed the pump, the bolt bound condition gave no room to make adjustments in the shaft gap even though coupling hubs could be adjusted.

4.6 CASE STUDY 5 – SUCTION PRESS ROLL INTERNAL BEARING FAILURE

FAILURE EVENT

The internal bearing on a suction press roll on a paper machine was experiencing chronic infant mortality Level 2/3 failures. The bearing was showing an inner race defect on vibration analysis within 6 months. The failure was identified prior to the Level 4 failure and did not cause any downtime cost impact.

FAILURE MODE

- Spalled inner race on one inboard side, as shown in Figure 4.7.
- Single spall mark on an inboard side, apparent surface deformation from impact damage – brinelling

FIGURE 4.7 – Bearing spall damage

POSSIBLE CAUSES

- Lubrication deficiency
- Static corrosion
- High thrust loading
- Impact damage (brinelling)

ANALYSIS

The bearing was lubricated with the correct lubricant. The lubricant was also in very good condition at the time of the failure. Static corrosion can cause roller damage on raceways but typically covers multiple rollers. Once spalling or pitting begins it will spread through the race. One of the failures had spalling over a large area but only on one side of the rolling elements. If it were corrosion, there would have been indications on the other side of the bearing as well.

This internal bearing generally only has radial loading. The failure was severe considering the run time. There was also loading indicated on the other side of the roller path which was very close to the loading on the spalled side. This information led the investigation to look closer at the position of the bearing and loading at various points of service from the roll conditions.

The bearing failure mode looked like impacting damage or brinelling. One of the failures only had a single spall and the rest of the bearing rolling surface looked very good. It could be seen on the single spall where the roller indented the inner ring. The spalled bearing had a point where a single impact was made. That bearing just ran longer for the spall to spread which somewhat covered up the failure origination point. The more spalled bearing had a thrust indentation on the guide flange which was at the same point where the roller impacted the inner ring. Now this looks like the same failure mode, just with one being operated a little further into the life cycle.

With the cause being determined that impact damage was the primary failure mode, what has caused the impact damage? Shipping damage was ruled out as all new bearings are inspected in the shop before installation. The analysis revealed that the damaged zone and the operating load zone for the bearing were not in the same area. Operating issues such as load and lubrication appeared not to be related to the failure since that portion of the bearing looked good. The investigation now focused on roll assembly.

During roll assembly, the suction box was pushed into the internal bearing which was already mounted into the roll head. This roll was a very large roll and required a large amount of force to push it into the bearing. The fork truck used to push the box was a little small for the job. The fork truck had to build up speed to get the box into the bearing. The speed and resulting impact on the bearing caused brinelling damage to the bearing.

ROOT CAUSE

The root cause of failure was impact damage during suction box installation where the box was hammering the internal bearing causing brinelling damage, which started the spalling. The latent root cause was poor procedure and tools for roll assembly.

POSSIBLE ACTION ITEMS

- Modify the roll assembly procedure for the suction box to eliminate bearing damage. A different method for getting the suction box into the internal bearing was investigated.

SUMMARY

The bearing's fate was set before it ever had any run cycles on it. One of the keys to solving this failure was having multiple failed bearings to investigate and properly identifying the failure mode. This led to further conversations with maintenance crews about how roll assembly was accomplished to find the root cause. Once the new roll assembly method was used the internal bearing failures never returned. The suction box was made with a taper and slide shoes for the installation but the issue was the box impact to the bearing.

4.7 CASE STUDY 6 – EXHAUST FAN BEARING FAILURE

FAILURE EVENT

An exhaust fan on a tissue machine was experiencing chronic bearing failures. All failures were Level 2/3 failures. The repairs were typically scheduled on a machine outage or the fan could be taken down and the bearings replaced on the run. The bearings would only last 2–3 months, and the failures had been going on for as long as the machine had been operating. Other failures also occurred around the fan system such as on the ductwork, fan housing, shaft, belt drive, and motor. While the cost of the failures was only ~$15,000 annually for material and labor, the interruption to planned work was costly. Also prior to a more extensive RCFA, a project was in the capital plan to replace the fan assembly for a cost of ~$400K.

The fan is a mist exhaust fan on the tissue machine, which evacuates moisture from key parts of the machine. While fan bearings are not expected to last forever (L10 life estimation was calculated at 4.5 years), bearing failures 4–5 times a year is excessive and had to be solved.

INITIAL FAILURE MODE

Below is the first attempt at solving the bearing failures.

- The first failure analysis showed a skewed roller path (3mm difference) on the bearing which indicated misalignment. The bearing also had a 360° load zone on the outer race, as shown in Figure 4.8.

INITIAL ANALYSIS

Following the failure mode observed, the following were possible causes identified – misalignment, bearing installation, lubrication, and belt drive assembly.

The skewed roller path in the bearing outer race indicated a bearing alignment issue even though it was not believed to be the root cause of failure. The fan side bearing is a self-aligning double row ball bearing that can handle some misalignment so it was not a major issue. It still showed that the fan and bearing assembly was not square.

The second big issue from the first failure was the 360-degree load zone on the bearing. It seemed like the bearing was overmounted (too much bearing RIC removed) causing excessive bearing loading. Since it is a ball bearing

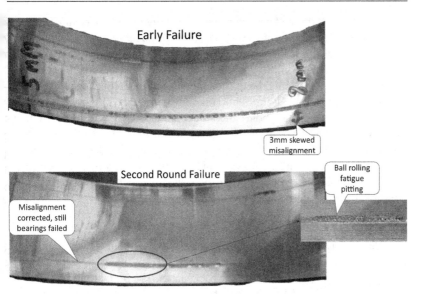

FIGURE 4.8   – Fan side self-aligning ball bearing outer race damage

on a tapered sleeve, installation had to be by the angular drive-up method. There is no way to verify RIC similar to a spherical bearing. The investigation revealed that there was not an established installation procedure and technicians were basically guessing at the installation. A normal load zone on a bearing is around 120° when properly mounted. This mounting induced a large stress on the bearing, which greatly contributed to the failure. Another observation was that the fan side bearing sometimes had the bearing housing mounted on different base holes, which left the fan more overhung and resulting in an increased bearing load.

At the time of the failure there appeared to be plenty of quality grease in the bearing and housing. The bearings had been operating around 180–190°F (82–87.8°C) with 460cst grease. Several different methods of viscosity calculations revealed the viscosity should be between 68–150cst between the ball bearing and spherical bearing applications. Due to availability and simplicity, a standard 220cst synthetic grease was selected for both bearings. The grease tag information on the fan called for 1.4oz (40g) of grease every 9500 hours. Greasing calculations showed about 1.3oz (36g) every 4–6 weeks. The last observation on lubrication was that the fan side ball bearing housing had the grease fitting on the W33 lube groove port location. Since this was a self-aligning ball bearing there was no W33 lube groove on the bearing so no lube could get to the bearing. The grease fitting needed to be moved to the side of the bearing housing.

The fan was driven by a belt drive, so installation was very important due to the resulting belt force on the bearing assembly. High belt loads can not only cause high bearing loads which can cause significant bearing life reduction but also shaft deflection causing other damaging high vibration. From the L10 bearing formula, doubling the load on the bearing can reduce the life by a factor of 10 so precision belt installation is very important.

Like bearing installations, the belt drive assembly was basically being done without any detailed tension information or tools. It was not known where the drive was being set. The spherical bearing on the belt drive side was not experiencing failures, it was the fan side bearing that had all the failures but the resultant load from the belt drive can still affect the life of the fan side bearing, but it would not have had a 360° load zone.

ROOT CAUSE FROM 1ST ANALYSIS

The first analysis basically tried to address the as-found failure modes and basic issues of the fan bearing design. Several key items were found to be potential root causes or contributing root causes. The initial root cause was the incorrect installation of the bearing by removing too much internal clearance. Contributing causes such as incorrect lubrication, misalignment and belt drive installation were also noted from the initial analysis.

1ST ANALYSIS ACTION ITEMS

- Change grease from SHC 460 to SHC 220 and change lube frequency to every 2 months.
- Establish an installation procedure for fan and fan bearings including detailed bearing installation instructions and belt drive tensioning.
- Move the grease fitting on the fan side ball bearing to the side of the bearing housing.

FAILURE EVENT 2

There was hope that the failures would go away after some good findings from the initial investigation, but they did not and the failures returned. Bearing life on some occasions was somewhat better, with a few extending to 4–6 months but nothing beyond 6 months, so maybe doubling the bearing life at best but still chronic bearing failure persisted. Round two RCFA commenced.

FAILURE MODE 2ND ROUND OF FAILURES

- The fan end of the self-aligning ball bearing had fatigue spalling and pitting on both rows of the outer race. This was like the previous failure, but the load zone had no misalignment. The bearing load zone was very good (around 120°, as it should be after precision installation from the applied angular drive-up method).

• There was a typical high overall vibration between .6–1.0 in/s (15–25mm/s) especially in 1X frequency indicating possible unbalance. It varied in magnitude depending on bearing issues and the amount of fan buildup before and after cleaning.

POSSIBLE CAUSES

For the second round of RCFA, the analysis had to have a different approach. This can be common on many failures, particularly on complicated systems. Chronic failures may also qualify if there has never been any RCFA completed. There is no reason to go complicated on what may be an obvious and easy root cause, but this chronic issue was not one that was obvious. The possible causes are listed below for the second attempt RCFA.

• Fan operating point
• System issues – ducts, dampers, valves, mist bank
• Baseplate issues
• Fan components – fan wheel, inlet cone, shaft, housing, etc.
• Unbalance
• Resonance

ANALYSIS

The fan is an overhung centrifugal fan with pillow block bearings and a belt drive. The fan is driven by a 250HP (186kW), 1800rpm motor. The belt drive slows the fan speed down to 1313 rpm. The fan wheel had an original G2.5 balance. The fan was originally sized for 64,908 CFM@10.6" H2O (110,300 m3/hr@2575Pa). The fan bearings are 2219EK self-aligning double row ball bearings and 22219EK spherical bearings. The basic fan layout is shown in Figure 4.9.

The fan operating point was checked with field measurements using a pitot tube and compared with the original design. It showed that the fan was running much more throttled than the original design. The fan was operating at ~14.5" (368mm) H2O total pressure compared to the original design of about 10.6" (269mm) H2O. Looking at the fan curve, this correlated to a ~9,000 CFM (255m^3/min) lower air flow. A previous air flow study was found on the fan and system, but in the RCFA it was discovered that they used the wrong fan curve so some of that study was not very helpful.

So, what was causing the fan to operate at the throttled condition? After a system walkdown, it was found that one of six suction valves was shut off. A review of historical documents showed that ducts had previously had some plugging from machine carryover fiber. Ducts were inspected on an outage and cleaned although no significant blockage was found. The fan discharge damper

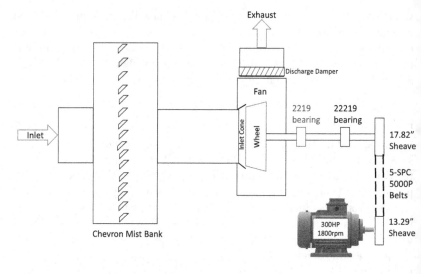

FIGURE 4.9   – Fan system general layout

was inspected and found open even though some of the linkage system was not in good condition. It basically stayed open all the time and never operated much.

The other fan system that was inspected was the mist bank which was on the fan suction. This mist bank had two rows of chevrons that were meant to collect dust and keep it off the fan wheel. Dust buildup on the fan wheel causes balance issues and frequent cleaning. The mist banks were both plugged significantly. This was observed on fan shutdown before an inspection as well. When the fan was shut off the change in pressure caused the ductwork to buckle. The higher pressure created recirculation in the fan which added to the overall vibration.

While the system was down, a close inspection of the fan components was completed. There was evidence of some shaft issues from fan housing rub and some fretting where a bearing sleeve was installed. Observation of the fan wheel showed some visual runout at the end of the fan wheel. The inlet cone had some cracking. The inlet cone and fan inlet did not fit properly and left too much gap and potential air recirculation. The wheel also had some buildup which needed cleaning.

Since one of the main failure modes was 1X vibration, the unbalance had to be investigated further. This was the normal condition monitoring indicator for cleaning the fan and it would drop slightly after cleaning but was still very high and at an unacceptable magnitude. One observation was that there were old balance weights all around the fan wheel. During an outage, all the old balance weights were removed, and a fan rebalance was completed. This improved the fan balance but, again, did not drop the vibration down to acceptable levels nor solve the bearing failures.

Comparing the original maximum speed and the operating speed, they are within a few rpm of each other. This was a red flag in the investigation. From looking at the belt drive in the first analysis, some drive sheave wear was noted. Drive sheave wear can affect actual running speeds as the belts seat lower down in a sheave which could have made the actual speeds on this fan well above the maximum speed. The measured run speed was 1,330 rpm instead of 1,313 rpm.

Fan cleaning had been inconsistent in frequency and quality. Partial cleaning of the fan was shown to increase unbalance. Letting buildup get too high would also lead to poor balance operating conditions.

The baseplate was inspected and in general was in good condition but looking at the span of the structural members, the support across seemed questionable. As a trial, wedges were installed along the long-supported span. Vibration dropped around 50% so the stiffness change affected the amplitude significantly. At this point resonance began to be suspected. Resonance magnifies an existing vibration when they collide at the same frequency. Resonance also is very destructive and the weak component in this fan assembly is the fan end ball bearing, since it is point loaded with the overhung fan wheel.

Resonance testing was conducted by the vibration analyst. One of the most prominent resonance points was at 1,440 rpm (24Hz). To prevent most impacts from resonance it is recommended to operate 15–20% away from any significant resonance frequencies. With that operating margin the fan should operate outside the speed range of 1,224–1,656rpm (20.4–27.6Hz). The fan is operating at 1,330 rpm (22.16Hz) which means the resonance greatly affects the fan vibration.

To change the resonance or natural frequency there are two variables that can be changed – mass and system stiffness. While the stiffness or mass of the baseplate support system could be changed, the resonance is on the rotor assembly which is not easily changed. The forcing function to change is the speed so it doesn't operate at the natural frequency. The easy variable to change, especially with a belt drive, is the speed of the fan. From fan performance information the fan is already being operated below target design so slowing down should not impact fan function within the system. The fan is already running at maximum speed, so we don't want any increase. A speed increase results in the drive power cubed, which we may not have.

$$f = \frac{1}{2\pi}\sqrt{\frac{k}{m}}$$

Where f = resonance frequency
       k = stiffness
       m = mass

ROOT CAUSE

The root cause of fan bearing failure was determined to be that the fan was operating with severe resonance vibration. The operating speed (1,330 rpm) was in the resonance frequency range (1,224–1,656rpm) with the resonance peak at 1,440rpm. The latent root cause is fan system design. Improper maintenance and operation of other fan systems contributed to overall fan reliability.

There were many contributing causes to bearing failure and high vibration, in addition to fan resonance, which included the following.

- The fan operating point was to the left of the design causing recirculation vibration.
- There were fan wheel to inlet cone fit and gap issues due to out of round on both pieces, causing air recirculation.
- There was pluggage on the mist bank on the fan suction ductwork and this was not being cleaned.
- There was unbalance of the fan wheel as a result of too long between cleaning and the poor quality of cleaning.

POSSIBLE ACTION ITEMS

- Replace the fan rotating assembly (wheel, shaft) with new. The existing shaft had some minor damage. The fan wheel had some small cracking from all the years of high vibration operation.
- Add an additional fan inspection and fan cleaning door on the housing.
- Change the frequency of fan cleaning and mist bank cleaning to monthly instead of quarterly. Add PM to CMMS for fan cleaning.
- Repair and stiffen the damaged/cracked ductwork around the fan.
- Modify the fan belt drive so that the fan speed is below 1224rpm to eliminate resonance impact on the fan bearings.

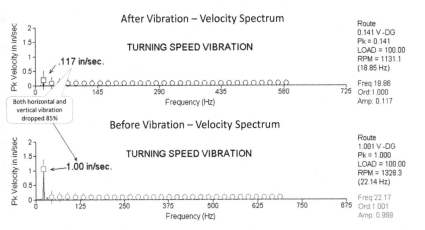

FIGURE 4.10 – Comparison vibration data

SUMMARY

Most manufacturers have not done extensive Finite Element Analysis or vibration testing on many of the products that are operating. The maximum speed was determined by the fan manufacturer for a reason. Field installation issues can also contribute to a piece of equipment's resonance vibration conditions. Many systems on this fan had been neglected for many years without knowing the impact on overall fan operation and reliability. There were so many needs to address that this RCFA took several iterations to address total reliability improvement. As with any chronic RCFA, the proof is seeing the corrections made and monitoring the results. All the contributing causes were corrected but high vibration continued as did the short bearing life. Once the fan speed change was completed an 85% drop in vibration occurred, as shown in Figure 4.10. Consequently, the bearings had no failures over the next 5 years of operation. The large capital project to replace the fan and system was taken off the spending list, and this all because of an RCFA that found the root causes of the failure which were corrected.

Key Factors in Pump System Failure

5

5.1 FUNDAMENTALS OF PUMP FAILURE

There are two main kinds of pumps in industry: rotodynamic (centrifugal) and positive displacement. For this focused topic discussion, pumps will refer to centrifugal pumps.

Pump failure can be difficult to solve unless there is a complete understanding of pump systems and the interaction with the mechanical pump. Mechanical pump failure can have mechanical origins or hydraulic (pump system) origins. Hydraulic or pump system issues can many times be the root cause of mechanical failures even though it may not be immediate but occurs over time. These can begin as Level 1 pre-failure conditions and progress to Level 2 and 3 and finally Level 4 functional failure.

Mechanical failures can also have their own mechanical-related root causes. Pumps have several components which typically can experience some level of failure. A study in one paper mill revealed the failure breakdown of these components, as broken down below.

- 41% bearings
- 21% impeller
- 16% couplings – a variety of failure modes – lubrication, misalignment, etc.
- 15% sealing – includes mechanical seals, packing mostly
- 7% shaft – includes shaft sleeves with main shafts

Many pump bearings are chosen to operate for 40,000hrs. For a 247 operation, this would be around 4.5 years which is in line with many studies. Low reliability pumps may last less than 12 months, and high reliability pump systems can operate 10 years with reliable operation. Pump bearings can have several failure modes. Typical bearing fatigue can occur if some other condition doesn't get them first. Corrosion on pump bearings can occur when water gets into pump bearing housings. Lubrication issues can stem from many causes such as contamination, an insufficient lube level or an incorrect lubricant. A 68 centistoke (cst) viscosity lube is a typical lube for most pumps. High speed pumps may be 46 or 32cst viscosity.

End suction pumps will have a radial bearing and a thrust bearing. The amount of thrust and radial loads depends on numerous factors of which many are directly related to the pump operating point, pump design and installation.

Pump impeller failure modes typically do not involve Level 4 loss of function-type failures. Most impeller failure is Level 2 or 3 where the failure mode may be wearing or pitting. Many of the pitting-type failure modes may be due to cavitation which can result in micro jets of high pressure impacting on impeller vanes. Severe impeller wear can cause a functional failure as the pump fails to produce enough flow but is a gradual condition.

Couplings typically have two major failure causes. Every coupling can have negative effects from misalignment. The misalignment can also adversely affect the bearing loads, shaft and sealing reliability. If the coupling is a lubricated coupling, then poor lubrication can also be a major factor in coupling failures.

Pumps typically have either a mechanical seal or packing although some pumps may use a dynamic seal. Mechanical seals are vulnerable to any operating conditions which may result in shaft deflection, which opens the seal faces. Seal flush conditions may also be a major factor. Packing requires some routine adjustments to keep functioning properly and must be periodically replaced.

A pump shaft can fail by fracture (fatigue or overload) which would be a Level 4 failure. Pump shaft deflection can occur due to a pump operating point that produces high radial loads. Pump and shaft design also will affect the shaft deflection. The L3/D4 shaft stiffness ratio for a pump design is an indicator of how much the shaft will deflect. Shaft deflection is just a Level 1 pre-failure condition but can lead to seal failure and bearing failure. Pump shafts can have other failures regarding shaft sizing and bearing fits. These can be Level 1, 2 or 3 types of failures and can lead to Level 4 functional failure if the bearing spins on the shaft.

Most pump failures have some element of pump system and hydraulic elements to the root causes. Most of the component failures are also closely related to the hydraulic operation of the pump. The pump operating point, or interaction with the pump system, will determine the magnitude and type of forces acting on the pump. Not only will this affect bearing life and shaft deflection but it will also affect vibration and cavitation. When thinking about a Level 4 functional failure of a pump where it fails to deliver the flow in a system, many times there is a system issue that is the root cause and not the pump that has failed. This is where pump and system interaction must be understood and analyzed.

When looking at a pump and system curve, the pump will operate where the system curve intersects the pump curve. There are many things that can be determined by looking at the pump system with regard to pump reliability and the potential failure issues that can result. Looking at Figure 5.1, there are several points to consider on the pump curve. For the best pump reliability, the pump should operate in the preferred operating region (POR) which, for low specific speed pumps, is -30% to +20% of the best efficiency point (BEP). The POR is focused mostly on efficiency and with that comes reliability, which has been set by the Hydraulic Institute (HI). The allowable operating region (AOR) is set by pump manufacturers and gives the range of operating limits for the specific pump.

The further away from the BEP and particularly outside of the POR, a pump has many negative effects. To the right of the BEP, cavitation can become an issue as the net positive suction head required (NPSHR) increases. Cavitation can increase vibration and affect pump performance. Shaft deflection can also become an issue with high flows. To the left of the BEP, cavitation from recirculation can become a problem. As radial shaft loads increase, shaft deflection can cause sealing issues and increased bearing loads. Extreme left

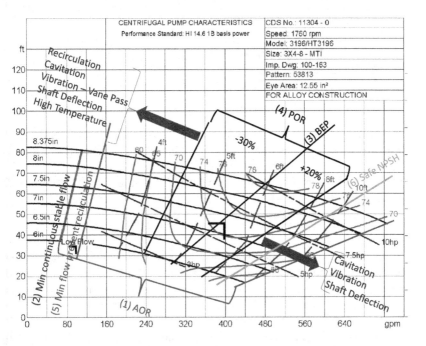

FIGURE 5.1 – Pump curve operating points

of the BEP and an increase in fluid temperature can result as the pump energy is built up on the recirculating fluid and pump efficiency decreases drastically. Pump operating conditions such as recirculation, cavitation or shaft deflection all increase vibration. The vibration energy can be a root cause of, or a contributing cause to, many of the failure modes around bearing failure, seal failure, impeller failure and shaft failure. A failure analysis of a pump failure that does not include an evaluation of the pump system and pump operating points is many times an incomplete analysis. To understand the pump operating conditions is to understand the reliability and possible failure modes for the operating pump. The following case studies will show some examples of pump failure and some of these interactions.

5.2 CASE STUDY 1 – WHITE WATER PUMP FAILURE

FAILURE EVENT

This failure was a chronic failure where the pump had experienced several different types of failures.

- Level 1 failure – high vibration, vane pass
- Level 2 failure – impeller pitting
- Level 3 failure – sealing (packing)
- Level 4 failure – bearings

The pump had a mean time between failure of ~7–9 months. Some of the time the pump would be identified through predictive maintenance that a failure was imminent but other times it came quickly. The pump always had a very high vibration even after installing a new pump rotating assembly. The annual cost of the failures was ~$20,000.

FAILURE MODE

While there were many different failure modes that would occur, the bearing failures seemed to be the main final issue that led to pump functional failure and premature scheduled replacements. Initial failure investigations did not reveal any bearing installation or lubrication issues with the bearings. High vibration was a constant pump issue. There was a full history of work orders since machine start-up, with this pump having high vibration and short mean time between failure (MTBF). It was a historic chronic failure pump with poor reliability.

The vibration was high on the pump regardless of having a new pump installed or having run for a period of time. Overall vibration trends had the pump around .5 in/s (12.7mm/s) and would spike near .7-.8 in/s (17.8–20.3mm/s) when the pump was nearing a Level 4 functional failure. High vibration is known to decrease bearing life significantly and this was suspected as the main cause of failures. The bearings were getting beaten to death. So, what was the root cause of the vibration and what could be done about it?

For initial root cause brainstorming by area teams, the failure mode was kept at a high level at first so as to not leave out any major root causes. The possible cause below was part of the initial failure analysis by area teams. Later the investigation and analysis focused on the detailed failure modes.

POSSIBLE CAUSES

The possible causes for the high vibration are listed below.

- Misalignment: The pump was aligned numerous times to <.002" (.051mm) offset and <.0006"/in (.015mm/mm) angularity which satisfies the standard ISO alignment targets for an 1800rpm machine. It was discovered in the analysis that the machine had a .005" (.127mm) potential thermal growth misalignment due to the growth of the motor and pump. This was also accounted for in alignment corrections once realized, even doing hot alignment, but the high vibration Level 1 pre-failure condition remained unchanged.
- Unbalance: Standard impeller balance for many centrifugal pumps is G6.3, as was the case for this pump. To improve operation, a new impeller balanced to G1.0 was installed many times as well as the correct key length for shafts and couplings. Once impeller damage occurs the G1.0 balance is quickly erased. The vibration signature did not have a dominant 1X vibration due to imbalance so the high vibration condition was not affected by impeller balance quality.
- Shaft Runout: This was also checked, and all passed acceptable runout at <.002".
- Pipe Stress: This was verified and was not a contributor to the pump operation or vibration.
- Foundation: This was found to be in good shape. The width and mass of the foundation was found to be a little low compared to the rule of thumb, but this would only affect the dampening of the vibration and would not be the forcing function or source of the vibration itself.
- Loose bolts or components: All fasteners were operating in good, preloaded condition and all other components were at pump manufacturer specifications when a pump was replaced. Some fit areas were slightly out of specification but were replaced with no impact on pump performance.

All these things were possible causes of high vibration, but they were not really failure mode-based. Several were corrected but they had no legitimate connection to the actual failure mode and chronic failures continued.

FAILURE MODE (2ND ROUND)

Later failure analysis took a much closer look at the failure modes on high vibration. The vibration had lots of flow turbulence and cavitation and a high vane pass amplitude. Ultrasound readings at the pump casing also revealed high levels of cavitation noise. The vibration signature is shown in Figure 5.2.

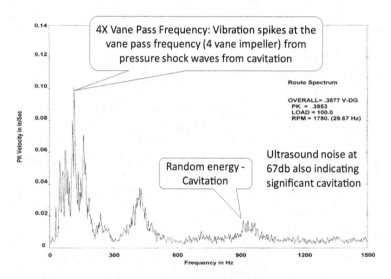

FIGURE 5.2 – Typical pump vibration spectrum

The failure mode of the pump impeller was the reoccurring pitting damage found every time a new impeller was installed. Figure 5.3 shows typical impeller damage found on the pump. The impeller had severe pitting on each of the four impeller vanes. Pitting was on the suction side of the impeller vanes, indicating a suction cavitation issue versus a discharge cavitation issue. The high impacting energy on the impeller vanes would be a reason for the vane pass vibration spike.

REVISED FAILURE MODE

• Suction side pump cavitation – there was severe pitting on all four impeller vanes near the impeller eye on the suction side of the impeller.

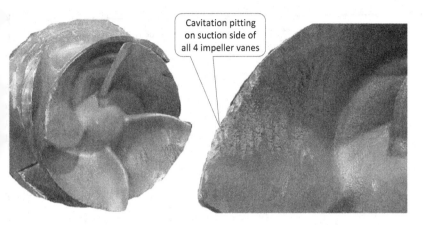

FIGURE 5.3 – Pump impeller damage A) 4 vane impeller B) cavitation pitting damage on impeller vanes

REVISED POSSIBLE CAUSES

1. Pump operating outside the POR and AOR
2. Insufficient Net Positive Suction Head Available (NPSHA)

ANALYSIS – DESIGN BASELINE

The pump is a 12x10x17 size pump with a 15.5" (393.7mm), 4 vane impeller. The current design point was 7000gpm@102ft (26,498 lpm@31.1m) at 79% efficiency. The pump driver is a 300HP (224kW), 1800rpm, 2300V electric motor. The original pump was sized for 6000gpm when the original machine had a 3000fpm design speed. A machine speed-up adjusted this pump to a design point of 7000gpm (26,498 lpm). There are work orders going back three decades since the machine was first started up, noting impeller erosion, changing rotating assemblies, a repacked pump, alignment work, and various repeated maintenance tasks.

POSSIBLE CAUSE 1 – PUMP OPERATING
OUTSIDE THE POR AND AOR

Installing pressure gauges and checking motor loads, the pump operating point was verified. The pump operating point showed it was operating at 5500–6000gpm (20,820–22,713 lpm) at 150–160ft (45–49m) as a stable process system. Figure 5.4 shows the operating point and design point of which both are within the POR for this pump. The pump has a rather large AOR and the POR is typically a much tighter target. While the machine speed-up slightly altered the pump design point, it was not significant to the pump operation. It was concluded that the pump operating point was not really a factor in the pump failure.

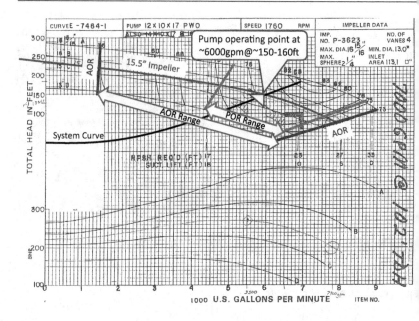

FIGURE 5.4 – Pump curve and system curves with operating points

POSSIBLE CAUSE 2 – INSUFFICIENT NPSHA

There must be a certain amount of NPSHA to suppress cavitation in any pump system application. It is noted that the NPSHA must be greater than the NPSHR. The pump curve NPSHR is around 21ft (6.4m) for the operating point. So, what are the possible causes where the NPSHA is insufficient? This is a good place to apply the physics of failure. The equation for the NPSHA is shown below. The NPSHA is an absolute pressure.

$$NPSHA = P1 + Hs - Hvp - Hf$$

Where NPSHA = Net Positive Suction Head, ft
P1 = Pressure on liquid surface, absolute pressure, ft
Hs = Static suction head of liquid level in tank
Hvp = Vapor pressure of liquid, ft
Hf = Friction loss of fluid in suction piping to impeller eye, ft

So, let's step through these variables in the physics of the NPSHA as part of the analysis to solve the failure mode of cavitation. The tank is an atmospheric tank with atmospheric pressure on the surface of the white water so 14.7 psi (101kPa) is around 33.9ft (10.3m) for P1. The tank level is typically around 10ft (3m) for Hs. The vapor pressure for the white water at operating temperature is 1.6 psi or 3.9ft (1.2m) for Hvp. The theoretical friction loss, Hf, was around 1.9ft (.6m). So, using the NPSHA equation, the NPSHA = 38.1ft (11.6m). Looking at the pump curve in Figure 5.4, the NPSHR is around 21ft (6.4m). The NPSHR for the analysis is at the actual operating point not the design point, which is different. The initial conclusion would appear to be that there is no cavitation concern as NPSHA of 38.1ft > NPSHR of 21ft so cavitation should be suppressed; however, why is there cavitation?

There is one NPSHA variable that must be verified – the friction loss. If the suction line had a blockage that isn't supposed to be there, it could drastically reduce the NPSHA to the point where cavitation may be a big issue. So, comparing the theoretical suction pressure (4.2ft) at the pump inlet to the measured suction pressure (installed pressure gauge at 4.6ft) revealed that there was less than .4ft difference in the two pressures. The actual pressure was slightly higher, so this confirmed that there was not a major blockage in the pump suction. This was also verified on an outage where the tank and suction line visual inspection was completed.

Back to the other conclusion from initial NPSH analysis. The NPSHR from the curve is not the NPSH where no cavitation is occurring. The curve NPSHR is the NPSH3 which means it is the NPSH where the pump head is reduced by 3% due to cavitation. So, cavitation is occurring at this NPSHR value. There must be some NPSH margin above the NPSHR or NPSH3 to fully suppress damaging cavitation on some pump systems. There has been much debate in hydraulic circles regarding the NPSH and cavitation prevention. While there are great insights by many methods and standards, this case study will show some results from a standard that was used at the time of the failure.

The ANSI/HI 9.6.1–1998 version defining suction energy and the NPSH margin was the method used to analyze this failure and many other pump system failures. Some of the elements are suction specific speed and suction energy. Equations for both are given below.

$$S = \frac{N\sqrt{Q}}{NPSHR^{.75}}$$

Where S = Suction specific speed
N = Pump speed, rpm
Q = BEP Flow, gpm
$NPSHR$ = BEP NPSH3, ft

$$SE = De*N*S*SG$$

Where SE = Suction energy
De = Impeller eye diameter, in (can estimate by .9*suction diameter)
N = Pump speed, rpm
S = Suction specific speed
SG = Specific gravity

For this pump, S = 13,756 and SE = 267,408,488. SE above 240x10^6 is considered a very high suction energy pump. The NPSH margin for this condition is 2.5. The new NPSHR is the NPSH3 times the NPSH margin so the NPSHR = 52.5ft (16m). Now the NPSHA at 38.1ft (11.6m) < NPSHR at 52.5ft (16m) so cavitation is a serious problem. A lower SE pump may only have an NPSH margin of 1.2–1.5 which would still have NPSHA > NPSHR. Note the SE ends up being a square of the speed, so speed is a major factor.

When this was learned on this pump, analysis was done on all the other process pumps, which found that the other very high suction energy pumps also had very high vibration, poor reliability and cavitation issues. Interestingly, the same exact pump operating at 1200rpm in other applications had over 75% less vibration and failures were not common. This is two more examples of the effective use of method of differences in analysis.

One other interesting finding on this pump regarding the design is that the rotating assembly on this size of pump was the largest size pump it was used on. This frame is the same power end that is used on an 8x4x17 pump, which handles 1000–1800gpm (3785–6813 lpm) versus the 12x10x17, which handles 4000–8000gpm (15,141–30,283 lpm). That meant the pump had the same shaft and bearings that were able to handle much bigger loads.

ROOT CAUSES

The physical root cause for this chronic failure was insufficient NPSHA to suppress damaging cavitation. The latent root cause is pump and system design, particularly around undefined pump selection standards on the NPSH margin.

ACTION ITEMS

There were many investigative action items from the initial failure investigations, such as correcting misalignment, thermal growth, key length, and impeller balance, none of which solved the chronic failures. In this failure analysis and situation, much is pointed back to the original design. There are not really any short-term actions to eliminate the cavitation issue and operate the process with the existing equipment. We could not affect the vapor pressure or pressure in the tank or friction loss. The suction head improvement was very small. Reducing suction energy and pump speed was not an option as the process needs could not be met with those changes. The main solution is a new pump for the system with much improved NPSH, SE and cavitation characteristics. There were some other action items discovered during the RCFA that fall under short-term actions.

Below are the remaining action items after the root cause was identified.

- Short term – Fix broken and missing pipe hangers.
- Short term – Install back pressure valve and pressure gauge for stuffing box on the seal water system.
- Short term – Increase the level in the white water tank to increase the NPSHA. This would only be 4–5ft so only a modest improvement but not enough to solve the problem.
- Short term – Reduce the impeller diameter to 15.25" (387mm). This will not solve the cavitation issue but will save some energy. Several of the system control valves were operating less than 30% open.
- Long term – Investigate a project to install a new pump for the whitewater system which satisfies the NPSH requirements (specify a lower suction energy pump, lower speed).
- Share RCFA learnings with predictive maintenance technicians to help identify other pumps with similar issues and pre-failure conditions.
- Revise pump sizing and specification standards to consider the NPSH margins on future capital projects.

SUMMARY

The pump and system utilized the best available technology and knowledge at the time of original installation in the 1970s. Later advancements in NPSH application knowledge allowed this lifelong chronic failure issue to be solved. All the attempts at precision maintenance for many years were all good activities but did nothing to address the root cause of failure for this pump system.

Follow-up on this failure after the action items were completed confirmed the accuracy of the analysis in finding the root cause. A new pump was installed to satisfy the hydraulic conditions and process demand. The new pump operated at 900rpm and had a suction energy of 141×10^6 which was nearly a 50% reduction. The resulting NPSHR was 11ft (3.4m) compared to the NPSHA of 38.1ft (11.6m). Cavitation was virtually eliminated, and the new pump ran for over 10 years with no maintenance until business conditions shut the machine down. The overall vibration dropped from ~.5-.7 in/s (12.7–17.8mm/s) to.08 in/s (2mm/s). Another example of failure mode-focused failure analysis.

5.3 CASE STUDY 2 – STOCK DILUTION PUMP FAILURE

EVENT

A dilution pump experiences loss of system pressure, as reported by operators, and experiences pump casing leaks from pitting. Due to pump wear the pump loses performance until the pressure setpoint can't be met. The pump and casing are changed each time to restore performance. Also, the pressure control recirculation loop experiences severe pipe vibration and cracking from normal operation.

FAILURE MODES

1. Pump casing pitting from cavitation damage to the integral casing suction wear plate, as shown in Figure 5.5.
2. Pressure recirculation control valve pipe vibration and cracking.

Suction wear plate pitted very bad in 2 locations (near cutwater on casing)

FIGURE 5.5 – Dilution pump casing pitting damage

POSSIBLE CAUSES FAILURE MODE 1

- Casing erosion/corrosion – incorrect material for pump casing
- Defective material from manufacturer
- Insufficient NPSHA to suppress cavitation
- Impeller rub

- Abrasive material in fluid
- Operating pump outside of the AOR or POR

ANALYSIS

For the design baseline, the pump is an 8x10x16 end suction pump with a design point of 3100 gpm@130ft (11,735 lpm@39.6m) and the NPSHR = 23ft (7m). The pump is driven by a 200HP (149kW), 1800rpm fixed speed motor. The pump was designed with an original 13.25" (336.6mm) impeller. The pump system is a dilution system for a pulp process and feeds many different dilution points with many control valve loops. A basic process design is shown in Figure 5.6.

While the pump casing had some flash iron oxide (rusting) present on the surface, that was not a primary failure mode. The fluid was white water at an acceptable Ph of 7.3 for the cast iron material (6–9 Ph). The white water had no abrasive material. There were no indications of any impeller rubs. The pump casing pitting damage was chronic and there had been many casing replacements. The material hardness was verified to be within the target (196–228Bhn) for the cast iron material.

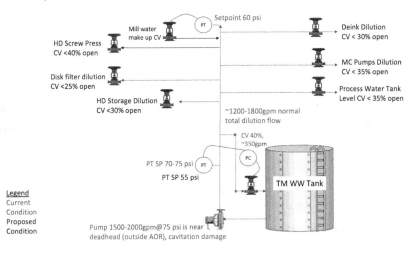

FIGURE 5.6   – Process design for dilution pump system.

Pump vibration data showed some vane pass vibration and vane pass harmonics but not much classic cavitation. The vibration levels ranged from .14 to .3 in/s (3.6–7.6mm/s) which was above the vibration limits per Hydraulic Institute 9.6.4 for end suction, foot mounted pumps above 100BHP (74.6kW) which is .18 in/s (4.6mm/s). Impellers with G1.0 and G6.3 balance specification had been installed on the pump with no measurable impact on pump vibration or reliability.

The pump curve NPSH3 = 23ft (7m) and with an NPSH margin of 1.5 yielded an NPSHR = 35ft (10.7m). The NPSHA was calculated at 41ft (12.5m).

NPSHA>NPSHR so pump cavitation was not a significant cause of the pump failure. The pump impeller showed no signs of any significant cavitation either.

The pump impeller size was verified on an outage as 13.6" (345.4mm). It was discovered that the impeller was enlarged to get more pressure from the pump since the casing wear was affecting performance. A 13.6" impeller would give full curve runout with the existing motor.

The existing pump has a BEP of 4400gpm (16,656 lpm) with a POR between 3080–5280gpm (11,659–19,987 lpm) and an AOR between 1800–6200gpm (6,813.7–23,470 lpm). Data was taken to verify the pump operating points. The motor load was estimated by measuring the motor current and voltage. The power factor and efficiency were taken from motor load tables. Using the formula below, the motor BHP was found to be 141HP (105kW).

$$BHP = 1.732*I*V*pf*\eta \, / \, 746$$

Where BHP = motor brake horsepower
I = motor current, amps
V = motor voltage, volts
pf = power factor
η = motor efficiency

The pump flow was estimated by calculating the pump total dynamic head (TDH) and known impeller on the pump curve. The pump TDH, using the measured pressure and tank levels, was determined to be ~162ft (49.4m). The pump curve would be around a 1500gpm (5,678 lpm) flow on the pump, as shown in Figure 5.7. This was also validated by using the pump BHP formula below. The BHP of motor and at pump is the same so rearranging the formula with a known TDH and efficiency resulted in a flow ~1500gpm. The operating point from two different sets of measurements is very close so confidence in knowing the actual operating point is high.

$$BHP = \frac{Q*H*SG}{3960*\eta}$$

$$Q = \frac{BHP*\eta*3960}{H*SG}$$

Where BHP = pump brake horsepower
Q = pump flow, gpm
H = pump TDH, ft
η = pump efficiency
SG = fluid specific gravity

The pump is operating so far outside of the POR and AOR that recirculation within the pump is very high. The recirculation-type cavitation is pitting the cast iron integrated casing wear plate. The impeller is stainless and withstands this recirculation energy without any damage. Looking back

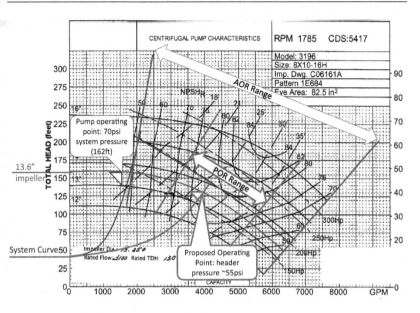

FIGURE 5.7 – Pump curve with operating points.

at the process diagram in Figure 5.5, all the control valves in the system were operating below 40% open and most were below 30% open. The control valves should be operating in the 40–60% open range for good control and reliability. Operators were operating the header pressure at a very high level. One reason for this is that the mill water makeup control valve pressure setpoint was too high. This kept the mill water makeup valve from using expensive mill water. However, whenever these two pressures were adjusted the pressure control valve loop off the header would open the control valve 100% which increased recirculation to the point where pipe vibration caused piping failures (failure mode #2).

The pump system as it was would either run control valves in poor positions or the pump in a poor operating point. The existing pump was simply oversized for the process as it was currently being operated. Either a new properly sized pump must be installed or an additional demand found for the existing pump so that the operating point can be within the POR and can eliminate the recirculation cavitation.

ROOT CAUSE

The physical root cause of failure is operating the pump outside of the AOR and POR. Low flow demand (high system-operating pressure) caused recirculation cavitation damage on the pump casing. The latent root cause is process changes from the original design to cause the pump to operate outside the AOR. There was not a management of change process where review was completed looking at the impact of modifications to the process.

POSSIBLE ACTION ITEMS

- Short term – Upgrade casing material to 316SS or CD4 to improve wear resistance.
- Short term – Install a smaller 13" impeller on the pump to reduce dilution header pressure to 55 psi (379kPa).
- Short term – Add a management of change process for any future modifications from original designs.
- Long term – Add additional flow to the pump system (~1500gpm). If process solution can't be found, add larger recirculation line back to the tank and install a larger control valve.
- Long term – Install a smaller pump on the system to match system demand (1200–1800gpm).
- Short term – Lower the header pressure setpoint to 50 psi.
- Long term – Investigate variable frequency drive (VFD) for a new pump and eliminate the pressure recirculation valve. At the low flow demand condition, saves ~$23,000 annually.

SUMMARY

The pump system operated for several years from original start-up with few issues. For the past 12 years of operation, pump failures have been common, which occurred after process changes took the hydraulic load off this pump system. Adding an additional 1500gpm of process flow to the current pump system allowed the pump to reduce recirculation and begin operating within the POR, which eliminated the pump failures.

5.4 CASE STUDY 3 – LOOP 1 DILUTION PUMP FAILURE

EVENT

A dilution pump operated until the pump casing would begin to leak and spray dilution water into the surrounding area. A new pump casing would be installed and operated until it began to leak in the same manner again. A new pump casing would last ~2.5 years before this Level 3 failure of pump casing would occur. Historical work records went back to the original machine start-up and noted the same issues. This chronic issue cost ~$20,000 every 2 years.

FAILURE MODES

1. Primary failure (Level 3): Pump casing pitting around suction wear plate area of pump casing until fluid leaks to atmosphere, as shown in Figure 5.8.
2. Secondary failure (Level 2): Pitting of stuffing box cover behind the impeller. This failure mode typically would not leak. This failure mode was figured to have the same root cause as the primary failure mode.

FIGURE 5.8 – Pump casing pitting damage on suction wear plate area.

POSSIBLE CAUSES

- Impeller rubbing
- Abrasive material in process fluid
- Corrosion on pump casing from process conditions and materials
- Air entrainment, submergence issue on pump suction
- Operating outside of POR/AOR – recirculation or cavitation

ANALYSIS

The pump is an 8x10x15 end suction pump with a 13.5" (343mm) impeller driven by a 150HP (112kW), 1780rpm VFD motor. The design point for the pump is 2600gpm@160ft (9,842 lpm@48.8) with NPSH3=17ft (5.2m). The dilution system controls dilution for the coarse screens, cleaner system, high- and low-density systems, a heat exchanger and water tank.

The normal operating condition typically had the VFD operating at 100% to maintain header pressure at a 68 psi (468.8kPa) setpoint. This left the operating point for the pump near 2750gpm@153ft (10,410 lpm@46.6m). The motor load also confirmed the operating point. All the dilution system control valves were operating in a range of 25–60% open. Other condition monitoring data was checked, such as pump vibration, and under operating conditions it was running very smooth showing no signs of cavitation or flow turbulence issues. Very low vane pass vibration was present but typical for this application.

The tank level feeding the pump normally ran around a normal level of 10ft. Looking at the pump using an NPSH margin of 1.5 at the operating condition, the NPSHR = 25ft (7.6m). The NPSHA was 30ft (9.1m) so NPSHA>NPSHR which eliminated cavitation concerns at the normal operating condition. It was becoming apparent that the problem condition and root cause was an intermittent process condition or event.

The impeller was verified to not be rubbing on the casing and there was no abrasive material in the dilution water. Casing corrosion was ruled out due to close inspection of the casing with heavy cavitation like pitting. The only corrosion was the quick corrosion on the surface of the pitting. The original casing material was ductile iron and had a suitable Ph range of 6–9. The dilution water tank Ph was verified to be steady at 7.3.

Air entrainment will not typically cause severe pitting damage, but it was easy to verify. This system could be vulnerable to air entrainment with the short tank having an overflow of ~13ft (4m). Air entrainment from the tank level being operated too low was evaluated by calculating the minimum submergence to suppress a vortex in the pump suction. The formula for submergence is given below. The minimum submergence was shown to be around 50 inches (1.3m) while the minimum operating tank level never dropped below 96≈inches (2.4m), so air entrainment was not an issue.

$$S = D + \frac{.574*D}{Q^{1.5}}$$

Where S = minimum submergence of suction line, inches
D = suction line diameter, inches
Q = actual flow, gpm

The pump appeared to be operating within the AOR and POR so no significant issues should exist, yet chronic pump casing failures were occurring. After installing a continuous vibration, monitoring and trending the header pressure, some spike events were observed. A random infrequent alternate operating point was discovered where a water tank fill valve would open 100%. This very low system head caused the pump to run well outside the AOR (1300–3600gpm) and POR (1925–3300gpm) when it operated at ~4000gpm. The NPSHR under this operating condition now was 42ft so NPSHR > NPSHA (42ft>30ft) and damaging cavitation resulted. Figure 5.9 shows the different operating points on the pump and system curves.

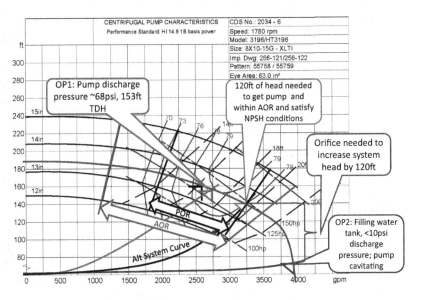

FIGURE 5.9 – Pump operating points and system curves.

ROOT CAUSE

The physical root cause was operating the pump outside the AOR. This intermittent operating condition created significant cavitation which damaged the pump casing leading to a very short service life. The latent root cause was that the system was not designed to keep the pump operating within the AOR during the alternate process condition.

POSSIBLE ACTION ITEMS

- Upgrade the casing material to 316SS to improve life with some cavitation conditions. This is not to address the root cause but was a first action when the root cause had not yet been identified.
- Add missing pipe hangers on pump piping which is relieving pipe stress to the pump. This was observed during the failure investigation but not related to the pump failure mode.
- Root Cause Action: Design an orifice plate on the water tank line so the operating point will be within the AOR and POR to eliminate cavitation during that process condition. The 10-year life cycle cost of correcting this chronic failure saved around $40,000.
- Investigate operating the dilution header at a lower pressure. This will provide an opportunity for energy savings in operation as control valves were mostly operating very closed and VFD was running 100% to maintain higher header pressure.

5.5 CASE STUDY 4 – REJECTS PUMP SYSTEM FAILURE

EVENT

This pump system Level 1 failure was a vibration issue which showed a large increase in vane pass vibration and component looseness. This component looseness resulted in a corrective work order to replace the #1 pump rotating assembly which was suspected on the final stage of failure. However, a closer review was done this time to make sure it was necessary. This pumping system had the #1 pump replaced a few times in the previous 4 years.

FAILURE MODE

1. Level 1 pre-failure condition: Vane pass vibration and component looseness on the pump rotating assembly.
2. Level 2 – hidden failure (leading to Level 3 minor failure of stuffing box): Abrasion wear on the stuffing box face behind the impeller and wear on the suction nozzle.

POSSIBLE CAUSES FAILURE MODE 1

- Bearing failure – defect, fit issue, wear
- Cavitation
- Operating pump outside the AOR

POSSIBLE CAUSES FAILURE MODE 2

- Impeller rubbing the stuffing box
- Corrosion of the stuffing box
- Insufficient material for process fluid conditions
- Cavitation – insufficient NPSHA
- Operating the pump outside the AOR

ANALYSIS

The system is a rejects water system with two identical pumps operating in parallel with a VFD to maintain the level in the rejects water tank. The two pumps are an 8x10x19 end suction pump with a 17.312" (439.7mm) diameter impeller driven by a 125HP (93kW), 1200rpm VFD motor. Full load amps (FLA) on motor is 155. At full speed the pump AOR range is 750–4000 gpm (2,839–15,142 lpm). The POR range is 2300–3900gpm (8,706–14,763 lpm).

For multiple failure modes, it would be typical to perform analysis on the highest level of failure first, but for this case study both possible causes were exhausted. The impeller was proven to not be rubbing against the stuffing box thus not causing wear or vibration issues. There was no significant corrosion on the pump components and the ductile iron material was a good choice for the application process conditions. Under normal operating conditions, there were no NPSH or cavitation risks for these pumps. Additional pump checks revealed that there was no significant bearing failure occurring on the pump. The only bearing issue left to check was the bearing fits but that would be after a pump change so other possible causes were explored first.

Evaluating the pump operating points, the DCS showed the pumps to be running at the same speed but with a different motor current. Now this did not make any sense in physics and was a red flag. Different pump impellers could cause this, but this was previously verified and both pumps had the same impeller. Digging into the DCS showed that the two motors did not reference the same FLA; however, this was not the main issue. A comparison between the data of the two pump motor loads is shown below. It is obvious #1 pump is not doing any work in the pump system.

- Pump #1 – DCS showed 36% load and 75% speed, 66 amps, 27 BHP (20.1kW) – referencing the correct FLA, % load should have been 43%

- Pump #2 – DCS showed 67% load and 75% speed, 131amps, 118 BHF (88kW) – referencing the correct FLA, % load should have been 85%.

The pump TDH was measured to be around 90ft (27.4m) which on the pump curve for 75% speed would have been ~2400gpm (9,085 lpm). The system piping discharge had a flowmeter which was reading ~1900gpm (7,192 lpm) during this process condition. Nothing was making sense until the speed was verified on each pump. The finding was that the pump speed on #2 pump was 100% and #1 pump was 75%. So now at 100% speed on #2 pump and 90ft (27.4) TDH, the pump may pump 3900gpm (14,763 lpm). Where is the extra flow of ~2000gpm (7,571 lpm) going?

With the speed difference, the system head is higher than the dead head pressure for #1 pump operating at 75% speed. The extra flow from #2 pump is recirculating backwards thru #1 pump since they are piped in parallel. The theoretical system curve shows where the pump system should operate at various flows. The alternate system curve is the lower head due to recirculation through the #1 pump short circuit back to the tank. The VFD for #2 pump was maxed out to control the tank level. Figure 5.10 shows the pump and system curves for actual operating conditions found during the failure investigation.

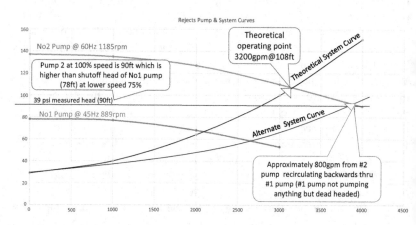

FIGURE 5.10 – Pump and system curves

ROOT CAUSE

The operating pressure of the system with #2 pump at 100% speed was higher than the dead head pressure of #1 pump operating at 75% speed. This caused #1 pump to be operating in a dead head condition. The result was pump recirculation that caused stuffing box wear. Not having the speed match on the pumps operating in parallel service caused #2 pump to carry the full load.

POSSIBLE ACTION ITEMS

- Adjust the pump speed so that each is operating at the same hydraulic load. Correct the communication between the VFD and DCS to reflect the correct drive speeds.
- Correct the FLA reference on each motor on the DCS readout.
- Train operators on parallel pump operation to ensure the same speed control.
- Upgrade the stuffing box material to 316SS on future replacement/repair of the worn stuffing box. This would be an insurance upgrade to improve component life and give more margin.

SUMMARY

With #1 pump operating in a dead head condition, the vane pass frequency is typically amplified, which is what was observed. The dead head condition and turbulence generated would cause lots of potential random forces which would also magnify bearing and shaft component looseness. Operating in this condition would also cause exponential wear as recirculated fluid rotates around the stuffing box and suction nozzle since the #1 pump has zero flow output, so the root cause was responsible for both failure modes.

Once the parallel pump system was corrected the operating conditions returned to normal levels. Not only were pre-failure conditions and Level 2/3 failures corrected but the overall system efficiency was improved greatly. Wasted recirculation flow is eliminated and both pumps now contribute to output flow. Overall, the root cause saved around $20,000 annually in operational energy costs. The as-found conditions were 145 BHP (108kW) to produce 1,900gpm (7,192 lpm) flow. After corrections the conditions were 100 BHP (75kW) to produce 2,300gpm (8,706 lpm) flow.

5.6 CASE STUDY 5 – SCREEN FEED PUMP FAILURE

FAILURE EVENT

Over a 9-month period the mechanical seal on a feed pump for a stock pressure screen experienced five mechanical seal Level 3 failures (seal leaking). The pump has been in operation for 14 years in the same process. Previous pump life was 3–4 years before any maintenance intervention. The mechanical seals began leaking in April, June twice, July and August.

FAILURE MODE

The pump had a couple of failure modes but the focus for the RCFA was on the seal failures.

1. Level 3 for the pump (Level 4 seal functional failure). Mechanical seal leaking: Stock was getting through the seal faces and into the seal springs. The seal faces were scored from abrasive material in the seal faces (paper stock from the process fluid).
2. Level 2 failure of the stuffing box: there was pitting on the stuffing box behind the impeller, as shown in Figure 5.11.

New stuffing box was taper bore open to product and no throat bushing

The original straight bore stuffing box had tight clearances and a throat bushing that increased shaft stiffness

Failure mode: erosion (abrasive wear) corrosion, concentrated at impeller diameter

Light brown flash rusting

FIGURE 5.11   – Stuffing box pitting

POSSIBLE CAUSES

Failure Mode 1 – Mechanical seal failures

- Mechanical seal defect
- Contaminated seal water supply
- Insufficient seal water supply
- Vibration
- Increased stuffing box pressure
- Shaft deflection – Operating outside of the AOR/POR

Failure Mode 2 – Stuffing box pitting

- Insufficient material for application
- Operating outside of the AOR/POR – Recirculation, cavitation, flow turbulence

ANALYSIS

The pump is a 16x16x19 end suction pump with a design point of 8450gpm@85ft (31,987 lpm@25.9m). The pump driver is a 250HP (186kW), 1200 rpm motor. The pressure screen that the pump feeds has a design feed pressure range of 20–75 psi (138–517kPa). This pump power end assembly is one of the larger units for the pump size (smallest size is 12x14x19) so there should be an ample design margin for the components. The largest size pump for this same sized power end (shaft, bearings, housing) is a 20x20x25 pump.

None of the mechanical seals were found to have any original defects. Many different seals have been installed, including one rebuilt seal (a new cartridge but the same seal gland) with the same results. The seals when changed early in the leakage showed seal faces in good shape but just had stock back in the spring seal load area. No evidence was discovered that indicated an issue with the seal itself.

The seal water system is a dedicated system with a pump and filter bank. The filters were found operating and water quality was not found to be a factor in the seal failures. The same seal water system feeds the entire operating area and pumps on the same system have had no seal failures. After the fifth failure a new issue was found with the seal water filters not backwashing and they were plugging as seal header pressure began to trend down for a week. This was not the root cause of the seal failure and was corrected when discovered during the investigation.

The seal water hookup found on the first failure was incorrectly hooked up on bottom port instead of the correct top port. Water still was in the seal but just above the shaft rotation. This has been seen on other applications and these not had any seal failures. This was corrected after the first seal was replaced and made no change in seal life or performance over the next four failures.

There had been no change in the vibration signature over many years of data. Only some small vane pass vibration and some raised floor vibration from slight cavitation and flow turbulence made up the otherwise low overall vibration. Since the long-term vibration trend was the same before and after the seal failures, this was not believed to be a major contributor. The detailed vibration data was taken on a monthly vibration round so this was during normal operating conditions.

The initial pump change was due to an oil leak 4 months prior to the seal failures. The rotating assembly was found with one bearing housing fit .0005" (.013mm) over target, which is typically not a big issue. Several new rotating assemblies were installed throughout the chronic seal failures but made no difference in seal life. Dimensional issues around the rotating assembly (the shaft total indicated runout, TIR, and bearing fits) were not considered to be a factor in the seal life.

There are many areas where stuffing box pressure can be increased. The formula for estimating stuffing box pressure is shown below for an impeller with balance holes. Higher suction head or pressure on the pump can increase the stuffing box pressure. This system is fed from an atmospheric open tank so the suction head is limited by the maximum tank level. The TDH is limited to what the pump can produce, which is higher under near shutoff conditions. The target seal water pressure feeding the stuffing box is typically 10–15 psi (69–103kPa) higher than the stuffing box pressure.

$$SBP = Hs + .1*TDH$$

Where SBP = stuffing box pressure, psi
$\quad\quad Hs$ = pump suction pressure, psi
$\quad\quad TDH$ = pump total dynamic head in psi

The pressure screen feed pump is started up under throttled valve conditions. So, for this system, the stuffing box pressure under the worst case conditions (near shutoff head) was 31 psi resulting in a target seal water pressure at the stuffing box of 41–45 psi (283–310kPa). Under normal operating pressure, the seal water pressure should be around 30 psi (207kPa).

After the fourth failure, a pressure gauge was installed to set and monitor the seal water pressure to the stuffing box. This is controlled by a manual ball valve and was found to be set at 15–20 psi (103–138kPa) which is well below where it should be during any operating point for the pump and meaning that the seal was not getting adequate seal water flush. This was at least a contributor to seal failure, but the seal started leaking about a month later, then briefly stopped leaking, then started leaking again. Previously the seal water had a needle valve flowmeter combination, but it plugged too often and was removed. This was typical on many of the seal water controls on most pumps.

The pressure in the stuffing box can also be influenced by the impeller clearance to the stuffing box as the back pump out vanes reduce the pressure. The impeller was found to be in very good condition but had about .010"

(.254mm) more clearance at that location than the new target. This measured clearance was after the new stuffing box was installed at the onset of these seal failures. The old stuffing box was pitting so badly that the clearance was actually more like.200" (5.1mm) with the old stuffing box so the impeller clearance actually improved at the time the seal chronic failures began.

The pump system showed to operate normally within the AOR (5,033–12,200gpm) and POR (6,900–11,700gpm), as shown in Figure 5.12. The TDH calculations seemed to have some error compared to the flowmeter measurements (TDH curve flow 5,700gpm versus flowmeter 8,600gpm). The shut-off head for the pump is 110ft (48 psi) yet there were many pressures above 50 psi and one as high as 70 psi. With system pressures that high, the pressure transmitter must have some calibration issues. The flowmeter mass balance appeared to be most accurate, which resulted in a 8,300gpm (31,419 lpm) pump flow which is very close to the system design of 8450gpm (31,987 lpm). The normal operation never got anywhere near that minimum flow of 5,033 gpm (19,052 lpm).

There was very light cavitation pitting on the low pressure (suction) side of the impeller indicating an NPSH deficiency at some point. Checking normal operating conditions and NPSH margins, the NPSHR = 46ft (14m) < NPSHA = 49.7 ft (15.1m) so not anything for significant cavitation. At 10,000gpm (37,854 lpm) the NPSHR and cavitation become more of a concern. The second failure mode did show pitting behind the impeller pump out vanes, which appeared to be hydraulic-related turbulence or recirculation-type cavitation. Also noted was wear near the cutwater location on the stuffing box, which again could be an indicator of recirculation-type damage.

During the troubleshooting process, continuous vibration sensors were placed on the pump and the trend showed spikes at system start-ups. On trending the system pressure, spikes also showed at start-ups. During a two-week timeframe, five pressure spikes > 60 psi showed at start-ups. These events were likely near dead head up to 2,000gpm and very far outside of the AOR for this pump (5,033gpm). The result would have been significant recirculation cavitation, vibration, and high radial loads resulting in shaft deflection. Looking back, one seal leaked immediately after start-up from the throttled start-up conditions.

While evaluating the pump system, the accepts control valve operated 35–38% open all the time. While thaat is in the allowable range (30–70%), it is outside the preferred operating range (40–60%) for a control valve. This operating point presented an opportunity for some energy savings. A smaller impeller could potentially lower the system pressure by 5 psi (34.5kPa) allowing the control valve to operate ~50% open saving ~30HP (22kW), yielding an $11,000 per year annual cost savings.

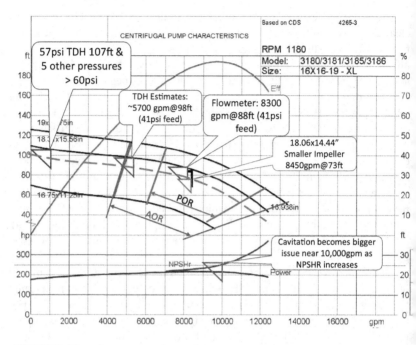

FIGURE 5.12 – Pump operating points

During the pump maintenance, when the failures first started, the following items were installed – a new pump rotating assembly, new stuffing box and new mechanical seal. The new component that was changed and constant during the chronic seal failures and all pump changes was the stuffing box. The new stuffing box installed was a taper bore style stuffing box and was different than the original stuffing box which was a straight bore stuffing box as shown in Figure 5.11. The original straight bore stuffing box also utilized a throat bushing. The straight bore stuffing box not only allowed the use of the bushing but also reduced the open area to process fluid access to the seal area and drastically reduced the stuffing box pressure in the seal area. The bushing acted as a third bearing that greatly increased shaft stiffness and supported the shaft outside the seal so the seal faces could remain parallel to each other under various operating conditions, such as outside the AOR operation. Shaft deflection would cause seal faces to open, allowing stock into the seal load springs.

ROOT CAUSE

- Physical root cause: The taperbore stuffing box created a more open area and increased the stuffing box pressure in the seal area to contaminate the seal. The taperbore stuffing box also removed the throat bushing which supported the shaft and seal geometry during various operating conditions where high shaft loads and shaft deflection occurred.

- Physical root cause: The pump was operating outside of the AOR during start-up or upset conditions causing high radial loads and shaft deflection opening the seal faces.
- Contributing cause: Insufficient seal water supply to the seal. Seal water pressure was found to be set too low based on estimated stuffing box pressure. Also, more seal water flush was needed for the taper bore stuffing box installed.
- Latent root cause: System modifications (different stuffing box and seal water supply, see Figure 5.11) and system start-up design conditions.

POSSIBLE ACTION ITEMS

A few items were corrected during the RCFA investigation but are still listed below to show the overall action taken.

- Correct the seal water hookup – correct port, pressure gauge, valve and pressure.
- Correct the seal water filter plugging. Consider a filter plug alarm for when auto backwashing is not functioning.
- Increase the tank level to 10ft to improve the NPSHA and reduce minor cavitation potential.
- Check the pressure transmitter calibration on the screen feed.
- Replace the existing cast iron taper bore stuffing box with a 316SS straight bore stuffing box and a new throat bushing like the original design.
- Investigate a smaller impeller for the pump since the system control valve is operating too throttled. Consider energy savings and reliability improvement for pump and valve operation.
- Improve the control valve response on start-up to limit system pressure spikes to below 50 psi.
- Share RCFA learnings with operators and maintenance technicians.

SUMMARY

The root cause of seal failures was a combination of the root causes because both conditions had to be present for the failures to occur. While the severe operation outside of the AOR on start-ups has gone on for years, it required the stuffing box change to affect seal operation. If the start-up conditions were corrected, potentially the taperbore stuffing box may not have been an issue. Confirmation through the method of differences showed that other pumps operate with taper bore stuffing boxes without seal failures, but they didn't operate outside the AOR.

When the new stuffing box was ordered, it was considered an upgrade or standard offering from the pump manufacturer so there were no concerns initially. There was also no understanding of why the start-up conditions were as severe as they were. Both were revealed because of the RCFA. The final proof was that after the stuffing box correction, the seal failures did not return

One other note was that one of the root causes of the seal failures (operating outside of the AOR) was also the main root cause for stuffing box pitting damage from recirculation cavitation. That damage had also been noted from maintenance records for years with no resolution to the root cause either. The stuffing box had just been replaced or patched every 5–6 years so finding the root causes had a far-reaching impact on pump reliability.

Key Factors in Mechanical Power Transmission Failure

6

6.1 FUNDAMENTALS OF MECHANICAL POWER TRANSMISSION FAILURE

Belt drives are one of the more common methods to transmit mechanical power in industry. Belt drives may be chosen for many purposes which include the following:

- Primarily to transmit drive power or generate torque to drive equipment.
- To change speeds of driven equipment.
- To allow an offset of driver when space limitations exist.
- To be the wear component of the drive system.

Belt drives can have other characteristics that can be advantageous, such as they are lubricant free, absorb shock loads, have a visual warning of failure, are low cost, and are easy to install. There can be many types of belt drives used on industrial equipment, but the V-belt and synchronous (timing) belt are the most common. V-belts transmit power by friction while synchronous belts transmit power by teeth engagement.

DOI: 10.1201/9781003248675-6

Belt drives must be designed so that there is sufficient belt drive capacity to drive the driven load with a proper service factor. Under-design or excessive over-design of a belt drive can lead to many problems when it comes to maintenance and component life. The service factor is also an important factor of belt drive design. V-belt selection can have many choices between single belts, a powerband, or notched belts (for small diameter sheaves). Contrary to common belief a powerband does not have a higher power rating than a normal V-belt. A notched belt can have a slightly higher power rating due to the better fit around the sheave and the friction generated.

Belt sheaves come in a variety of materials and construction. Different materials have different maximum design speeds. The maximum belt velocity for cast iron is 6,500fpm (1981mpm), 8,000fpm ((2,438mpm) for ductile iron and 10,000fpm (3,048mpm) for steel. Minimum sheave size is also something to watch for the smaller sheaves, which are typically on the driver (motor). Sheaves are installed on shafts by various methods such as single piece bored sheaves, taper lock (TL) bushings or Quick Detachable (QD) bushings.

Installation of belt drives to ensure full life has several elements. Alignment is where the sheaves and belts are aligned in the offset and angular planes. Alignment should be targeted as close as possible over the span. Many sources may list a maximum alignment of 1/32" per foot of center distance but with modern laser equipment much better alignment can and should be obtained. Consideration should be given as to the location of sheaves on the shafts such that the overhung loading is minimized. This will directly impact reaction bearing loads and bearing life. Due to variations in potential belt length, it is a best practice to use the same manufacture belts, so that lengths are as close as possible.

The sheave bushings have a minimum shaft size with which they should be installed. If the shaft is too small for the bore, it can crack the bushing by trying to clamp onto too small a shaft. Typically, it will crack at a keyway corner or another weak point.

Belt tension is one of the most important elements of belt installation. There are two types of tension – static and dynamic. Dynamic tension is the belt tension reaction due to the drive load inertia. When operating, the belt will have a tight side and a loose side depending on the direction of rotation of the driven load. For belt installation, the ideal tension is just enough belt tension to drive the load without slipping. There are many tools that can be used to calculate and set belt tension for an application. Using an engineered belt drive solution to determine belt tension is best since it will analyze the specific drive and not maximize belts used on the application, which will result in over tension. There are several methods to set belt tension, and each has limitations. A technician needs to understand all methods ranging from the belt deflection method and ultrasonic to the powerband multiplier.

Maintenance on belt drives involves most of the things listed previously and a few more items. Guarding issues at assembly can be problematic if there is not sufficient adjustment to account for belt drive tensioning, which affects shaft center distance changes. Don't use belt dressing to address a noise issue but instead find the root cause of the noise – misalignment, belt tension, worn components. Proper torque and assembly of sheave bushings is also very important. Checking for sheave wear and replacing are key maintenance issues. The small sheave is the one to pay special attention to since the contact friction area is so small, higher wear rates will result. These are typically on the motor sheave for slow-down applications.

Worn sheaves not only affect mechanical performance of components but also the function of the process equipment. When sheaves wear, the speed will change. When speed changes, the transmitted power will change at a cube of the speed change. If a powerband is used on a worn sheave, then the friction area is drastically reduced, and premature failure will result.

6.2 CASE STUDY 1 – AGITATOR BELT DRIVE FAILURE

FAILURE EVENT

A sludge tank agitator was experiencing chronic Level 4 failures over many years. Most of the time this was addressed by shift maintenance and was basically a run-to-failure maintenance strategy. The MTBF was less than 2 months and the mean time between work order was 1.1 months, costing ~$7,000 annually.

FAILURE MODE

There were many reasons for the work orders, such as belts burned, broken, or replaced (52%), Preventive Maintenance (PM) work orders (26%), blown fuses and troubleshooting (11%) and vibration (11%). The motor and drive system had experienced poor reliability for many years.

- The failure mode of interest was belts burned off, broken, and replaced.
- The motor had also run hot and had trip out events. An arc flash event in the starter cabinet occurred as the failure analysis was in progress.

POSSIBLE CAUSES

- Driven load does not match driver capability:
  - Excessive driven load – agitator setup – blade angle, blade size, speed, etc., process conditions – sludge consistency
  - Driver rated higher than drive system

- Under-designed belt drive system
- Deficient belt drive installation conditions.

ANALYSIS

The agitator is a size 30.20 with the belt drive system having a 5" (127mm) diameter sheave on the motor and a 19" (483mm) diameter sheave on the agitator. The original agitator design speed is 303 rpm. There are 6-3VX800 drive belts driving the agitator. The motor has a motor nameplate rating of 20HP (15kW), 1175rpm, 460V and 25.4 amps. The motor power factor is.805 with 92% efficiency.

The driven load from the process did not reveal any significant issues around consistency and tank levels. The agitator set-up was correct with the correct blades at a 15° angle. There was one interesting discovery on the agitator and motor speeds. The nameplate motor speed was 1175rpm which resulted in an agitator speed of 309rpm versus a design speed of 303rpm. The agitator driver had a rated speed of 1160rpm (higher slip motor). Using the affinity laws shown below, the power change would mean an additional 1.2HP (.9kW) for the drive not in the original design, which is likely not the root cause but a contributing cause adding drive load to the drive system.

$$P2 = P1\left(\frac{N2}{N1}\right)^3$$

Where P2 = Final power
      P1 = Initial power
      N1 = Initial speed
      N2 = Final speed

The motor was running hot up to 200°F (93°C) at times. The motor amperage was running 33.4amps which was well above FLA of 25.4. The motor voltage was also at 477V which was 17V above the nameplate but below the +10% typical allowed. However, that higher voltage translated into a higher load, especially with an already higher amp draw. The voltage unbalance was around.1% between phases which was acceptable. The actual BHP on the motor was 27HP (20kW) versus the motor nameplate of 20HP (15kW). It was no surprise that the motor ran very hot.

So why was the motor still operating with a 27HP (20kW) load? The motor overloads were found with an overload element that was good for up to 35.5amps which was larger than the correct one for a 20HP (15kW) motor. This likely occurred after motor trip issues. During the failure analysis, an arc flash event occurred in the starter cabinet. A loose connection was suspected as part of the issue in the arc flash event. The motor cable size and fuse sizes were verified to be sized to handle the amp load.

The drive belt system was evaluated and had a BHP rating of 24.4 (18.2kW) using some belt drive design software. The target motor BHP rating was 28HP (21kW) with a service factor of 1.4 so the belt drive was slightly deficient. So, under current conditions, the motor and belt drive system was operating in the service factor which means lots of failures.

Inspection of the belt drive loaded showed an operating temperature of 200–220°F. Drive belts should operate below 140°F for good performance and reliability. The belt drive was also very noisy, and a burnt rubber smell was noticed. Obviously, the belt drive was slipping causing excessive operating temperatures and very low drive efficiency. Looking at the belt drive motor sheave at 5" (127mm), it was observed to be slightly under the minimum National Electric Manufacturer Association diameter of 5.2" (132mm). The belt drive design was part of the original agitator manufacturer design and not much could be done to improve this part. Further inspection of the belt drive showed the motor sheave to be significantly worn, as shown in Figure 6.1.

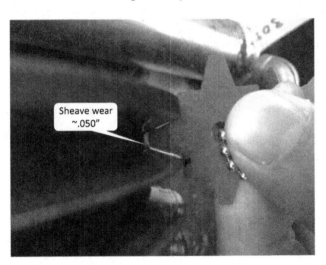

FIGURE 6.1   – Motor sheave wear

There were lots of work orders completing PM of what was being inspected or adjusted. The detail was very poor as even though the belt tension was on the sheet, there was no mention of how much and there were no details of the procedures or tools available to properly tension the belts. Discussions with maintenance in the area revealed no method to properly tension the belts at that point. Also not listed on the PM report was no alignment check, no check for worn components – sheaves, belts, bushing condition or bolts.

ROOT CAUSE

Insufficient belt drive capacity to drive the connected load was due to several factors:

- A worn drive sheave
- Low belt tension
- Some misalignment
- Belt drive design – the actual speed above design increased BHP demand, minimum sheave caused a low arc belt contact, low Service Factor (SF) on the drive belt design
- Wiring in the starter cabinet – potential wiring connection issues driving up the amp load

POSSIBLE ACTION ITEMS

- Investigate the redesigned belt drive. Due to agitator frame limitations, the option of drive belt redesign was not feasible
- Replace worn drive sheave
- Investigate belt upgrade with higher HP capacity and engineered tension to transmit a higher BHP and satisfy 1.4 SF
- Install new drive belts and sheave with precision installation on belt tension and alignment
- Add drive components to equipment bill of material (BOM)
- Purchase precision maintenance belt tension and alignment tools
- Complete belt drive training for maintenance technicians
- Build a belt drive installation worksheet for belt drive maintenance
- Repair the starter cabinet from arc flash event
- If above corrections do not reduce BHP then investigate 25HP drive for agitator or modify agitator (lower angle, remove blades) to reduce agitator HP demand

SUMMARY

After corrections were made, the belt temperature dropped from 220°F (104°C) to 127°F (53°C). Using the Arrhenius rule for every 18°F (8°C) increase in temperature, belt life is cut by 50%. The reduction in belt temperature should increase belt life by 5 times. The vibration dropped 50% after belt drive corrections. A new belt selection provided a potential 36BHP (27kW) capacity. There was no reason for the belt drive upgrade once the basic drive conditions were restored. The belt drive ran over 6 years without any corrective maintenance after these action items were completed.

6.3 CASE STUDY 2 – SCREEN BELT DRIVE SHEAVE FAILURE

FAILURE EVENT

The motor sheave on a primary screen belt drive experienced a Level 4 failure right after starting up from being down from a process control system crash. The sheave was not available in stock anywhere and the area was down for over 18 hours causing about $150,000 of financial impact. The drive sheave was temporarily welded back together to start up the process until a replacement sheave could be installed.

FAILURE MODE

- Primary failure – 14" (356mm) diameter 5-groove drive sheave for the pressure screen fractured in three even pieces at the TL screw positions as shown in Figure 6.2. Fracture appears to be brittle overload.
- Primary failure – Belts were burned off sheaves, sheave was hot. Powerband belts were all cut and split on the outer band.
- The outer flange on one side was fractured in a few places. This was presumed to be secondary failure after the sheave fractured when pieces hit the ground.

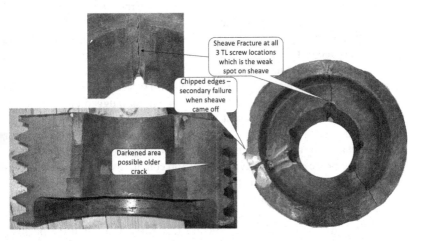

FIGURE 6.2   – Fractured drive sheave

- Defective drive sheave – casting defect, material
- Worn belt drive components – sheave, belts
- Overstressed drive sheave – excessive drive load, over-torqued sheave fasteners

ANALYSIS

The screen belt drive is driven by a 200HP (149kW), 1190rpm motor. The belt drive has a 14" (356mm) sheave on the motor and a 30" (762mm) sheave on the screen. The sheaves use a TL bushing and have one 5-groove 8V1600 powerband drive belt. The TL bushing used ¾-10x2 screws with a target torque of 200ft-lbs (271 N-m).

According to asset history, the sheave that failed appeared to have run for about 9 years. Infant mortality would not appear to be the failure pattern, in which case a significant manufacturing flaw might show up. There were no indications of any defects on the sheave upon close inspection.

The sheaves were checked for wear and the drive sheave was worn ~.030" (.762mm) in the grooves. Sheaves should be replaced at 1/32" (.031") (.787mm). Since this belt drive had a powerband on it, sheave wear was even more critical. The outer band on the belts was cut because of the worn sheave and poor V-belt friction on the sheave groove, which put more contact pressure on the outer band. Once the outer band is cut the result is a much lower belt tension to operate the belt drive and slipping belts occur at a much lower drive load. Belts slipping would also result in additional heat to the sheave which adds additional stress. Vibration had jumped up on the last reading which may have been because of some looseness in the belts as the condition began to deteriorate.

There are several potential sources for sheave stress. The rotational speed effect was verified to be within design limits. The maximum allowable rim speed for a block style gray cast iron sheave is 7,500 fpm (2286mpm). The actual rim speed of the drive sheave was 4,300 fpm (1311mpm).

A normal screen running load is 60% (530 ft-lbs) (719N-m) of full load rating of the motor driver (883ft-lbs) (1197N-m). The breakdown torque for the motor would be 1,324 ft-lbs (1795N-m) if the screen plugged and loaded up. When the screen begins to plug, it will go in a purge cycle to try and clean. This would happen well prior to a large event that would put the drive motor near a breakdown torque condition. This failure occurred during the start-up phase of the process area. As typical, start-up conditions are high-stress environments. On start-up the screen plugged four times, the sheave failed on that last plug. Belt drive stress was obviously higher during this failing condition.

The TL bushing screws were not torqued when installed. Typical practice is typically an impact tool to install fasteners. Being that the bushing and sheave are mounted on a tapered bore fit, the amount of screw preload will determine the mounting stress and fit. It is critical to get the fit correct to not overstress the drive components. The tapered portion of bushing had residual copper anti-seize on it. This was also a practice that had been observed previously where anti-seize was applied to all fit surfaces. This will drastically increase the amount of mounting sheave stress when screws are tightened. Original manufacturers state on installation procedures not to use any kind of lubricants on the tapered surface and that doing so can cause sheave fracture.

As per the mechanical power transmission association, the combined sheave stress should be less than 10% of the minimum tensile strength of the material. The reductions are due to endurance limit, residual stresses, and the safety factor.

ROOT CAUSE

Overstress of the drive sheave caused brittle overload, fracturing the sheave at the weak points (bushing fastener areas). There was a combination of stresses that contributed to the overstress condition, which include the following:

- Bushing installation error using anti-seize on TL bushing
- Likely over-torque of bushing screws using an impact wrench
- Thermal stress from burning belts off due to worn sheave on powerband from poor belt installation (tension) and worn sheave
- Increased drive load from potential plugged screen after process control downtime

POSSIBLE ACTION ITEMS

- Change worn drive sheave
- Correct BOM for belt drive parts for screen
- Establish belt drive installation procedure sheet with target belt tensions and sheave installation
- Purchase belt drive installation tools for alignment, sheave wear templates, and belt tension
- Complete belt drive training for maintenance technicians

SUMMARY

This failure was almost a perfect storm combination of conditions with the sheave wear, the overmounting of sheave, and the elevated drive load from the plugging screen. The sheave fractured at the weakest point where the screws are located. This failure revealed many deficiencies that had broad-reaching impact to many other critical belt drives. The true value of this RCFA went beyond the single event as many other belt drive installations were corrected after these findings.

Gear Drives

Gearboxes and gear drives are another very common method to transmit power in manufacturing processes. Gear drives may be chosen for many purposes which include the following:

- Primarily to transmit drive power or generate torque
- To change the speeds of driven equipment
- To change the rotation of shaft equipment from clockwise to counterclockwise

Some advantages of gear drives over belt drives can be improved efficiency, no slip, a longer life, and they transmit higher loads. Gearing has several standards – American Gear Manufacturing Association (AGMA), American Petroleum Institute (API) and International Standards Organization (ISO) which gear design and applications can be found in these standards. Gearing can come in a range of quality from AGMA Q3-Q15 (higher # is better). High-speed and high-power drives will often require higher-quality gearing. Noise and vibration may also be improved with higher-quality gearing.

Gear drives must be designed so that there is sufficient capacity to drive the driven load with the proper service factor. Gearbox design has changed since the late 1950s to where the power density has increased with improvements in materials. This has affected gearbox sizes, weight, operating temperatures, and numerous other areas. Gearboxes will have a mechanical rating and a thermal rating.

Gearing has many different types such as spur, helical, worm, planetary, hypoid, spiral bevel, etc. Helical gearing has a resulting thrust load from the gear reactions in which there is not only a thrust force magnitude but also a direction. The magnitude of the thrust load is determined by the transmitted load and helix angle of gearing. Gearing can be parallel shafting or right-angle shafting designs.

Simple gear trains (series gearing) can be single, double, triple, or quadruple reduction. Intermediate gearing only changes shaft direction. First and last gearing determines gear ratios for gearboxes. Compound gear trains utilize some parallel gearing where one shaft has multiple gears and gear meshes.

Gear lubrication is one critical element for avoiding gear failures. Gears have many challenges with lubrication. At the pitch line, gearing has a rolling contact and beyond the pitch line there is sliding resulting in boundary layer lubrication. A gearbox also has multiple lube conditions to deal with but only one lubricant to accomplish the job. For example, a triple or quadruple reduction gearbox is dealing with different loads and speeds from input to output. The bearings also deal with many different conditions throughout the gearbox. One lubricant must meet all those conditions. EP additives are typically recommended for gearboxes to improve lube film under these tough conditions.

Like any application, the lube viscosity is the most important property. Most gearbox manufacturers list the recommended lube viscosity on the name-plate along with other valuable specifications such as the lubricant volume. Many gearboxes will list the viscosity as an AGMA number between 0–15. An AGMA 5 is an ISO 220cst viscosity lubricant. Each number increases one ISO grade from there. An AGMA 5 would be a typical lubricant for most gearbox applications. A worm gearbox would use an AGMA 7 (460cst) lubricant many times but always consult the original equipment manufacturer. Some gearboxes may use a dipstick to check and verify oil volume/level. This must be verified while the unit is not operating due to changes in oil level while in operation.

Gearbox lubrication is typically either splash lubrication or pressure lubrication. Splash lubrication utilizes oil level and gears, picking up oil and slinging oil to bearings and gears. Pressure lubrication utilizes a pump distributing oil to lube points in the gearbox. Splash lubrication relies on the oil level being correct as a critical checkpoint. An oil dipstick may be utilized for verifying oil level and must have the gearbox down for an accurate assessment of oil level. An oil sight glass may also be used to verify oil levels. Running and down oil levels most of the time will not be the same so careful attention to detail on level monitoring. One advantage of pressure lubrication is the oil level is not quite as critical and there is the ability to filter the oil before application.

Lube oil cleanliness targets are typically at ISO 17/15/12 range for most gear applications. An often-overlooked filtration opportunity is air breathers where contaminants can be pulled into gearboxes. These can be particle- and moisture (desiccant)-type breathers and need to be the same micron rating as the oil filter used.

Gear failure modes can come from various categories. Surface fatigue (along the pitch line) can be in the form of pitting or spalling. Gear wear can be scoffing, scoring, abrasion or adhesion where boundary layer lubrication or contamination can be some of the causes. Plastic flow is when tooth material has deformed beyond yield and can be caused by overload and contributing to the cause of misalignment. Plastic flow is not a lubrication issue. Gear alignment must be good to attain full life and a healthy gear tooth contact pattern. Poor gear alignment can also cause edge chipping.

Tooth fracture failure modes can be from fatigue, overload (shock) or wear causing tooth thinning and leading to tooth fracture. Tooth fatigue is not as common a failure mode as it is mostly designed out of gearing. Due to very high load cycles, gearing must be designed so that at torque rating the stress is well below the endurance limit for infinite life. This is because of the high number of stress cycles a gear tooth can see in a short amount of run time.

6.4 CASE STUDY 3 – PAPER ROLL GEARBOX FAILURE

FAILURE EVENT

Several paper roll drive positions were experiencing chronic failures. New gearboxes were installed when failures occurred as it was not economical to rebuild from these failures. There were also some positions where oil leaks were an issue.

FAILURE MODE

There were multiple failure modes for the chronic failures which were severe gear teeth wear (Figure 6.3) and catastrophic bearing failure. Since new gearboxes replaced old ones, rebuild errors were eliminated from the list of possible causes. The failures were only occurring on the new style gearboxes, not on the original units.

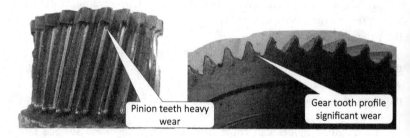

Pinion teeth heavy wear

Gear tooth profile significant wear

FIGURE 6.3   – Gear wear

POSSIBLE CAUSES

- Gearbox overload
- Electrical drive issues

- Misalignment
- Insufficient lubrication

ANALYSIS

The original gearboxes on the paper rolls were a Falk 2EZ1 with a 5HP (3.7kW) direct current motor drive. The Falk gearboxes became obsolete so an upgrade to a different Dodge model APG gearbox was necessary. The drive was also converted to a 10HP (7.5kW) variable frequency drive (VFD) on the gearbox change. A comparison of the catalog specifications of the two gearboxes is listed below.

- Original Falk Gearbox – 5HP (3.7kW) mechanical rating, 1.5 SF, ratio 2.791:1, 7.5 thermal HP (5.6kW) rating, 1217 in-lbs (137N-m) rating with SF
- Upgrade Dodge Gearbox – 21HP (16kW) mechanical rating, 1.0 SF, ratio 2.7:1, 10 thermal HP (7.5kW) rating, 2040 in-lbs (231N-m) rating with SF

The new gearboxes were obviously not undersized from a specification standpoint. The torque conditions were evaluated from multiple conditions such as the normal running load (NRL) and acceleration loads with no issues found. To reach a max torque rating of the new gearbox, the paper roll would have to be accelerated in about 19 seconds. The highest torque condition seen was about 761 in-lbs (86N-m) of acceleration torque. For 10HP (7.5kW), the rated torque is 1,229 in-lbs (139N-m). The NRL was very low, around 20% of rated load.

There were no drive tuning issues with any of the VFD components on any of the paper rolls. Alignment was also checked on all the rolls and no misalignments were found outside of targets. Laser alignment was done on every gearbox installation during conversions or changeouts.

Lubrication was the most suspected possible cause considering the damage found on the failures. The oil leaks were a red flag when compared to the original gearboxes, which did not have oil leaks. The original gearboxes also ran about 30–40°F (-1°C to 4°C) cooler than the new gearboxes although many of the new gearboxes were only running about 145–150°F (63–66°C). The amount of load on them at that temperature seemed elevated when compared to the original gearboxes with lower thermal HP ratings.

The original and new gearboxes were using a 220cst gear oil. Looking closer at the new gearbox specifications, it was calling for an AGMA 7 gear oil which is a 460cst viscosity. Likely during the conversion, the same lubricant was used as previously run.

Another critical issue discovered was that many of the new gearboxes had the wrong oil level. These small gearboxes have many different mounting positions in the machine. The design has many different plugs in the housing so oil levels can be changed depending on mounting positions. These units were mounted in horizontal, side, and inverted positions. Either the wrong ports were chosen or there was an incorrect oil level due to the failure of the sight glass on most of the new gearboxes. This was also the reason for oil leaks on some gearboxes.

ROOT CAUSE

The physical root cause of new gearbox failures was the incorrect lubricant selection and lube oil level.

POSSIBLE ACTION ITEMS

- Change oil in all new gearboxes to 460cst EP gear oil. Attach lube tags on all the gearboxes since original and new units are still in operation using different lubricants.
- Oil levels are determined by the mounting position they are installed.
- Provide some lube training on gearbox lubrication to lube technicians.
- Correct oil leakage from gearbox vents by attaching vents to correct locations and correct oil levels.

6.5 CASE STUDY 4 – COATING ROLL GEARBOX FAILURE

FAILURE EVENT

A machine coater has just experienced a paper break. After cleaning up and trying to start up the machine again, the #2 backing coating roll would not come up to speed. Further inspection revealed that the drive shaft from the gearbox was not turning. This failure on the backing roll drives had become chronic and random over many years with a gearbox lasting between 2 and 3 years. With only one spare gearbox for two positions, solving this chronic failure became very important.

FAILURE MODE

There were multiple failure modes during different failures so below is a list of some of the different failure modes that are also shown in Figure 6.4. The same root cause was believed to be behind all the failures.

- Primary failure – Fractured variable speed output shaft from the differential gearbox outside the bearing on the overhung right-angle gear. Most shaft fractures had a torsional and bending component of stress. Fracture appeared to be a fatigue-type fracture.

FIGURE 6.4   – Gear damage

- Primary failure – Spiral bevel gearing on the output shaft was fractured and chipped around the entire gearing.
- Secondary failure – Output shaft bearing housing was fractured. Housing is a cast housing and this appears to be overload failure.

POSSIBLE CAUSES

- Insufficient gear lubrication
- Gearbox rebuild errors
- Gear or shaft material issues
- Operating gearbox outside of design

ANALYSIS

The line-shaft driven coater machine is driven by two 400HP (298kW) direct current motors. The line shaft is connected to two size 5 Beloit differential gearboxes to separate dryer sections and two size 1 Beloit differential gearboxes connected to backing rolls, as shown in Figure 6.5. The backing roll size 1 differential gearboxes are rated at 75HP (56kW). Gearbox sizing for paper machine line shaft drives is typical for 200% startup torque for high inertia drive sections so the size 1 gearbox should be good for an intermittent torque of 150HP (112kW).

POSSIBLE CAUSES

The lubrication was an EP gear oil with a 220cst viscosity and was correct for the application. All the lube lines were in good condition and supplied

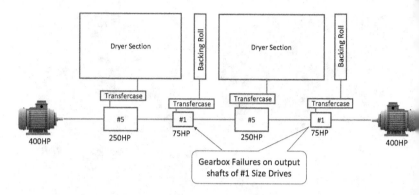

FIGURE 6.5 – Machine lineshaft drive train

lubricant to the gear nips and bearings on the output shafts that failed. The lube system is equipped with an internal oil pump which feeds all the lube points inside the gearbox. The pump feeds an oil tray and gravity disperses the oil to the lube points. The lube tray was found with some contamination but not restricting oil flows. Further inspection showed that there was no filtration system on the pressurized lube system, so any internal contamination was fed directly into critical components. While this did not directly cause this failure mode, it was a good learning to improve overall gearbox reliability.

The gearboxes had been rebuilt many times by a specialty gearbox repair shop. No deficiencies were found on any rebuilds. The failures did not include any bearing failures. Inspection and operating conditions (vibration, temperature) were all very good on all the gearboxes up to the time of catastrophic failure. The gearbox shaft and gear materials were found to be consistent with the specification design and the same as all other gearboxes used on the machines.

The NRL of the backing roll gearbox is around 22HP (16kW) which is well under the rated load of 75HP (56kW) and 200% intermittent acceleration load. Typical line shaft drives have clutches for each section between the differential gearboxes and the transfer case. This line shaft does not have any clutch or torque regulating device separating the drive sections. All machine sections in this design are directly connected by drive gearing. The analysis began looking for other drive conditions which may create high drive stress and utilized physics of failure to compare conditions.

The line shaft VFD has regeneration braking to stop the line shaft when a paper break occurs. The regen braking typically will stop the line shaft between 90–120 seconds. For drive analysis, an average machine speed of 2500fpm (762mpm) was used. Using the physical properties of the backing roll, the roll mass moment of inertia is 95,041 lbf-ft^2. The acceleration torque generated by the mass moment of inertia is given by the formula below.

$$T = \frac{WK^2 \Delta N}{307.2\Delta t}$$

Where T = torque in ft-lbs
  N = change in speed in rpm
    t = change in time for speed change in seconds
  WK² = mass moment of inertia

The torque is calculated at the backing roll inshaft drive. Typical acceleration time is around 227 seconds, which would equate to 259 ft-lbs (351N-m) torque. A typical regen braking deceleration time is 109 seconds, which is 539 ft-lbs (731N-m) torque. At 100% #1 differential gearbox rating, the drive torque at the backing roll inshaft is 2,073 ft-lbs (2,811N-m), as shown in Figure 6.6. Under normal operating conditions of NRL, acceleration and deceleration, there is no problem with operation versus design.

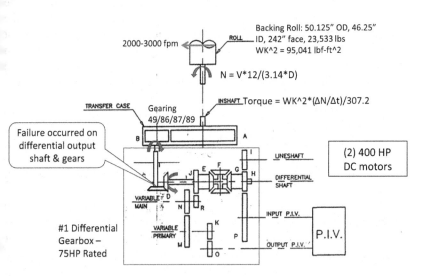

FIGURE 6.6   – Backing roll drive detail

One other operating condition to cause stress on the drive train is a sheet break on the backing roll. When this happens, the sheet can wrap the backing roll causing a deceleration of the backing roll. At 200% gearbox rating, the torque output at the backing roll would be 4,146 ft-lbs (5621N-m). Solving for the deceleration time in the equation for that torque reveals that if the roll stopped in 14.2 seconds then the gearbox would exceed the 200% intermittent load rating. Based on some operator observations, the backing roll can have a severe paper wrap which maximizes drive load in less than 14 seconds on occasions.

For the failure mode of overload on gear teeth, the high drive torque deflects the gear shaft causing poor gear meshing and even tip to tip teeth engagement. The overloaded output gear bearing housing failure is from overload, as the resultant force from the output shaft exceeds the design. The broken bearing housing on the shaft is evidence of a likely one-time event high-torque load. The shaft fractures may be from a single overload event or repeated stresses creating fatigue cracking until the final fracture zone on a subsequent stress event. The backing roll experiences many breaks and wraps during every shift.

ROOT CAUSE

The root cause is an excessive deceleration time creating a high drive torque which exceeds the gearbox design. Severe backing roll sheet wrap is the source of the high deceleration drive torque. Since the entire drive train and machine is geared together, the backing roll is trying to stop faster than the rest of the machine. The backing roll effectively becomes a brake for the rest of the machine whenever the backing roll deceleration time is lower than the rest of the machine deceleration, which includes two high inertia dryer sections. The weak link in the connected drive system is the output shaft of the size 1 differential gearbox.

POSSIBLE ACTION ITEMS

- Add gearbox drive torque to the historian data system for trending the drive system.
- Improve the sheet break detection system and coater head unloading to reduce the severity of backing roll sheet wraps.
- Investigate the clutch or overload system for backing roll drive to protect the gearbox during high load events.
- Investigate VFD for backing roll drives and remove the backing roll gear drive from the line shaft system.
- Modify lube oil systems on differential gearboxes by adding an external filtration system.

SUMMARY

To protect the small gearbox, a torque limiting device must be inserted. A standard clutch, similar to that found on typical line-shaft drives, will not fit between the differential and the transfer case so a torque-limiting coupling was selected instead of a VFD on the backing rolls, mostly due to project cost. Once the torque limiting coupling was installed, the differential gearbox failures were eliminated.

One last note is that the failures increased significantly after a coater head replacement, which occurred about 5 years earlier. While not the root cause, this was a contributing cause for the increase in the number of events. The backing roll size changed making clearances smaller between head and roll. Decreasing the time wraps would load up the drive. It was also believed that the head unloading was not as good as the previous design. These changes made sheet breaks more severe in wrapping the backing roll, which increased the drive torque back to the gearbox drive.

6.6 CASE STUDY 5 – PRESS ROLL GEARBOX FAILURE

FAILURE EVENT

The machine operators were doing rounds and smelled something burning. Upon inspection the press roll gearbox was found to be hot and on fire. The machine was shut down and the spare gearbox was installed, and normal operation resumed. No other issues were found other than the gearbox failure. The failure cost ~$80,000.

FAILURE MODE

The gearbox had several failure modes so understanding the failure mechanisms of each was critical. Being a catastrophic failure, the failure modes are identified as primary and secondary failures. A summary of failure modes is listed below and also shown in Figure 6.7.

- Primary failure – high-speed input shaft bearing catastrophic failure. The inner race appeared to have become very hot as rollers melted into the inner race and the shaft spun until the inner race was worn down (gone in some places). The outside of the rollers was flat to the outer race side, so they were stationary, not rolling, as failure progressed. The outer race did not have any spalling or normal-looking defects and was still in its housing. The cage was in relatively good shape from a wear standpoint as cage pockets looked pretty good. The cage was ripped apart from a severe torque lockup condition at final failure.
- Secondary failure 1 – high-speed bearing The seal cover was worn due to shaft contacting seal cover. The shaft contacted the seal cover due to bearing failure. Seal cover acted like a plain (sleeve) bearing on the shaft as high radial loads resulted by the secondary failure 2.

- Secondary failure 2 – high-speed input shaft fractured between coupling and bearing. Once the bearing failed, the thrust pushed the gearbox shaft into the motor shaft. With no bearing holding the shaft radially, the shaft began to walk around the motor shaft turning creating a large bending stress on the gearbox shaft until final fracture. Once the drive shaft broke there was no power driving the press roll.
- Secondary failure 3 – high-speed ODE shaft cover was fractured after DE bearing failure and the shaft thrusted away from the ODE. This created a large clearance on the ODE tapered roller bearing which resulted in the shaft hitting the guard.

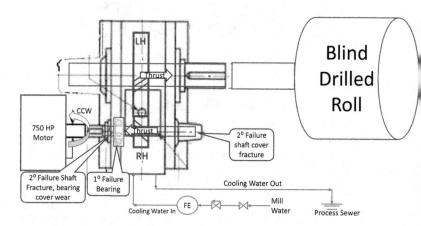

FIGURE 6.7 – Press roll gearbox

POSSIBLE CAUSES

- Insufficient bearing fit on drive end thrust bearing resulting in shaft spinning on the inner race initiating the catastrophic bearing failure.
- Insufficient lubrication causing increased friction. As bearing heats up, clearance is reduced and increased torque on bearing eventually leads to shaft spinning on inner race like lost shaft fit.

ANALYSIS

The gearbox is a Falk 385A single reduction unit with a ratio of 3.286:1. It is driven by a 750HP (559kW), 1800rpm VFD motor. It has a mechanical rating of 1263HP (942kW) with a 1.68 service factor. It has cooling tubes and is designed with splash lubrication. It uses tapered roller bearings on all bearing locations.

This failure is one of a chicken and egg scenario and is among the most difficult to solve. There are usually multiple strands of circumstantial evidence that lead to a root cause of failure. The bearing spun on the shaft, but did it lose fit or lose lubrication?

The target shaft fit from the gearbox manufacturer is .0015/.0025" T (.038–.036mm). The bearing fit was destroyed in the failure so there is no way to know what the fit was. The gearbox had run for almost 5 years, so it wasn't exactly a quick infant mortality failure. The shop which rebuilt the gearbox has closed and is no longer in business so going back there to retrieve rebuild records was not an option. One other element to this possible cause is what kind of operational load is on the gearbox? A higher load will work the bearing fit more. The actual motor load was around 250HP (186kW), so the drive is very lightly loaded with an actual SF of 5.

The gearbox vibration was very good and was unchanged leading up to failure. However, a bearing defect was detected but it was the bearing on the high-speed shaft on the ODE, not the DE, that failed. Significant fit issues may have shown some looseness or increased gear mesh, but neither was present in the vibration data. At this point, it is assumed there was not a major issue with the bearing fit.

The target lubricant for the gearbox is an AGMA 5 lubricant which is a 220cst EP gear oil. The gearbox was running a 220cst synthetic EP gear oil so no issues on correct lube. The amount of lubricant drained from the unit was 7–7.5 gallons (26–28 litres). The manual for this gearbox had 10 gallons (37.9 litres) as the target oil volume. Some oil was likely lost during the failure but based on the failure scene it was very little, so it appeared that the oil level was low by a few gallons. Based on the volume of the gearcase, 2 gallons (7.6 litres) would be around 1.75" (44.5mm) of oil level which would put the operating level well below the minimum oil level on the dipstick.

Looking at the dipsticks on the failed gearbox and the spare gearbox, there was a difference of about 1.25" (31.8mm) in dipstick length; however, upon further inspection and measuring the target oil level, this was the same on both dipsticks (only the length of dipstick was different). The dipstick is the only current method to check or verify oil level.

The bearing that failed was in the high-speed shaft with the thrust bearing taking most of the load (see Figure 6.7) so if there were a deficiency in lubrication this would be the bearing to fail first.

The other key condition monitoring data is the oil analysis on this gearbox. For over a year the gearbox quarterly oil analysis showed high water contamination from 1,500–24,000ppm (.15–2.4%). The oil samples also showed elevated iron 60–300ppm. The oil was changed several times and during investigation of the water source, it was determined that the water was coming from the cooling tubes which were leaking. The cooling water was shut off and the water contamination stopped in the last two quarters of oil samples; however, the iron wear elements increased significantly to 1,200–1,700ppm. The ODE bearing defect showed up about 2 months before the failure of the DE bearing.

After the cooling water was cut off, the oil was changed every 5 weeks during a scheduled outage except for the outage before the gearbox failed. The cooling water flowmeter was found to be set at 8gpm (30.3 lpm). The target cooling water flow is 2–5gpm (7.6–19 lpm). The gearbox manual even says cooling water flow >5gpm (19 lpm) may cause tube erosion and leaks. So, the root cause for the water contamination appears to be a cooling tube failure due to excess cooling water flow. Water contamination certainly affected the lubrication effectiveness, but the water was cleaned up in the prior 6 months before failure. Still, why was the oil level so low? First, more on the gearbox temperatures in operation and thermal HP.

While the temperature of the gearbox increased with cooling water off, it was still in the 180°F (82°C) range, which is below the AGMA sump tempera-ture of 200°F (93°C). This warranted a check on the thermal HP ratings of the gearbox. The thermal HP rating of a drive is the actual power in HP (without the SF) that a drive can transmit continually for 3 hours without overheat-ing. Overheating by AGMA is a sump temperature of 200°F (93°C) (assuming 100°F (37.8°C) ambient). The thermal HP rating is lower than the mechanical rating of a gearbox. The thermal HP should be higher than the connected HP. Several factors will reduce the catalog thermal HP such as ambient tempera-ture, altitude, air velocity, inlet water temperature and duty cycle. The thermal HP results are shown below.

- Gearbox Design: 750HP Motor, 1263 HP mechanical rating, 1.68 SF, Thermal HP rating 1009, Adjusted Minimum Thermal HP 639. Based on motor capacity, cooling tubes required for gearbox cooling.
- Gearbox Actual Conditions: 270HP Motor, 1263 HP mechanical rating, 4.9 SF, Thermal HP rating 1009, Adjusted Minimum Thermal HP 639. Based on actual load, could do without cooling tubes

While a preferred operating temperature of 150–160°F is desired, a higher temperature could be tolerated if it is below the 200°F. The lubricant is syn-thetic so it can also withstand a higher operating temperature. This gearbox did operate near 180°F in early summer as expected.

Looking closer at the gearbox cooling water in Figure 6.7, the cooling water exit was gravity to the process sewer. With the cooling water on, pressur-ized water leaked into the oil. With the cooling water off on the feed side, the oil could leak into the cooling tubes and run straight into the sewer. With no oil level indication, oil volume was lost over time. When the cooling tubes were valved out, the oil was changed at every outage (mostly due to higher operating temperatures) but with the outage before the failure, the oil was not changed.

The oil changes for 6 months were re-establishing the correct oil level even though it was not known that the oil level was low. After 10 weeks, the oil level reached a dangerous level as the failure occurred.

ROOT CAUSE

The root cause of the catastrophic bearing failure is a low oil level due to oil leaking out of the cooling tubes through the drain. The cooling tubes failed most likely due to high cooling water velocities through tubes for many years. Operating the gearbox for over a year with high water contamination could be a contributor to gearbox failure.

POSSIBLE ACTION ITEMS

- Adjust cooling water flow down to 2–5gpm on the gearbox.
- Re-check oil level in the new gearbox. During analysis it was discovered that not filling the oil troughs on initial fill can lower the oil volume by 10% (1 gallon in this case). This was not done on the initial fill.
- Investigate adding a needle valve to regulate water flow instead of an on/off ball valve which has difficulty regulating flow.
- Communicate to operating and maintenance technicians the cooling water requirements and temperature limits for gearboxes.
- Investigate adding an oil level sight glass on gearboxes.
- Investigate adding an access platform to the gearbox so that the dipstick can be accessed.
- Investigate other gearbox drives for thermal HP and the need for cooling water. Audit and adjust cooling water flows. If cooling water is cut off this 385A unit, a lower oil level is required, and a new dipstick or marking must be made.
- Investigate adding an annual gearbox inspection to the PM procedures.
- Investigate the need for a small oil dehydrator for small gearboxes or lube systems.
- Add oil analysis action items to the agenda for monthly predictive maintenance review.

SUMMARY

This RCFA may not have been as absolute in its findings for the root cause, but that is the reality of many failure analysis situations in most plants. Exact data is not always available. The analysis did reveal some very good findings and action items to prevent future failures and improve reliability and the bottom line. The gearbox audit and thermal HP evaluation showed that the other three gearboxes did not need cooling water. Annual savings was around 11 million gallons of water estimated at mill costs of $7,500.

Key Factors in Bolted Joint Failure

7

7.1 FUNDAMENTALS OF BOLT FAILURE

Fastener failures are very common in every industrial environment; however, there are many elements of fastener joints that are not fully understood by users. While there are many types of fasteners, we will concentrate on mostly the bolt- or screw-type fastener as this accounts for most of the fastened joints used on machines. Many use the terms "bolt" and "screw" interchangeably but there is a technical difference. A bolt has a nut as its counterpart while a screw is threaded into a blind hole that is tapped into the mating part. The fastener can be a bolt in one assembly and the same fastener can be a screw in another. As typically done in industry, the term bolt may be used in a generic sense for the rest of this discussion. Screw will be used for the specific application as referring to an actual screw, by definition.

While bolted joints can fail, there are usually some gross errors that have compiled to create the failure. This is partly because most bolted joints have a large service factor in the original design, especially when it comes to the actual bolts having a functional failure. A fastened joint has functionally failed when the two mating parts are no longer held together. Technically, if a joint is found with bolts still in it but the two mating parts have separated, then the joint has functionally failed.

The first requirement for a reliable bolted joint is an adequate design. The bolted joint must have sufficient bolt preload or joint clamp force to keep the two surfaces together from the design external loads. Some design standards will even call for a service factor of 5 for the bolted joint to handle the external loads (100% of external load to be 20% of joint clamp design). For the sake of bolt failure, we will assume that the designs are sufficient, as most designs are.

DOI: 10.1201/9781003248675-7

The large variables for why bolted joints fail are typically around assembly and installation conditions which reduce the performance of most bolted join failures. The rest of this introduction on bolted joints will discuss areas around installation that can cause bolted joint failures.

Many of the bolted joint issues are just fundamental ones with a few being more complicated. One of the first basic elements is to verify that the correct bolts, nuts, and washers are used on the joint. Correct means many things such as overall length, grip length, thread length, thread type (coarse or fine), metric or standard, grade (SAE 2,5,8, or metric 8.8, 10.9, 12.9), material (carbon, stainless or high temperature B7), head design (hex, socket) and many other basics. These may seem trivial, but many failures have resulted because some of these basic elements have been overlooked.

At assembly, is there good engagement of all bolted components? Thread engagement equals thread engagement factor times bolt diameter. A general rule of thumb for a thread engagement factor is 1–1.5. The thread engagement factor can be 1 for same or common materials (bolt and nut) and closer to 1.5 for softer materials such as aluminum or bronze. For a bolt (using a nut), the bolt should extend through the nut a few threads as the first few threads are not fully formed. Longer bolts will be made standard with an unthreaded shank portion. For a screw, the threaded portion should be in the shear plane of the joint or the joint will be loose. In this case the screw may get tight (reach a specified torque) as the bolt bottoms out against the shank, but the joint will be loose, not having any joint clamp load.

Another point to note about thread engagement and thread stress is that the stress on the threads is highest on the first thread and diminishes thereafter. Some studies show that 80% of the thread load is absorbed by the first three threads. The other threads are still important as they provide necessary service factor design. Also, when thread wear occurs on the first threads the stress gets distributed across the rest of the nut threads as load distribution shifts.

To properly assemble a bolted joint, the bolt must be tensioned or pre-loaded, which produces a clamp load resultant force on the joint. Preload and clamp force many times are used interchangeably as they are the same force but just from different perspectives – the bolt versus the joint. Bolt torque is the action on the bolt to produce the desired bolt preload which produces the joint clamp force. Bolt torque is not the goal; bolt preload and joint clamp load is the goal. Just torquing bolts is not good enough. The installer must control the torque conditions to establish the target bolt preload.

For properly designed bolted joints, the external load should stay within the compression zone of the joint when the bolts are properly preloaded. The compressive energy of the joint absorbs the external loads while keeping bolt loads low and the joint still in compression. In general, lubricants should be kept off the joint surfaces where lower friction may tend to loosen joints or put bolts in shear to handle the external loads.

An example of this would be a power transmission coupling flange. The amount of bolt preload (flange clamp force) required to transmit drive torque is proportional to the bolt circle, drive torque, and flange friction of the joint surfaces. So, if flange friction is reduced by lubricants, then a much higher clamp load would be required for the joint not to slip. Some technicians like to put grease on flanges to hold gaskets, not realizing that joint slip is likely in those conditions.

Two things happen when a bolt is preloaded, Fi, – the bolt stretches and the joint compresses. The magnitude of each depends on the joint stiffness, C. When an external load, P, is exerted on the joint, the joint compression decreases as it takes the load. The force on the bolt, Fb, also increases slightly. If the external load exceeds the joint compression, then the force in the bolt takes all the remaining external load as the joints have no compression and have separated. This can be seen in a joint diagram, as shown in Figure 7.1.

Most of the bolt torque is not converted to bolt preload. Only ~10–20% is converted to bolt preload. The other torque consumers are thread friction ~40%, and head friction ~40–50%. The basic bolt torque formula is shown below.

$$T = KDF$$

Where T = torque in ft-lbf
  D = bolt diameter in ft
  F = bolt preload or clamp load, lbf
  K = nut friction factor, K=.15 lubricated, K=.2 dry

The nut friction factor (K) can vary, depending on lubrication, and bolt finish so a specific nut factor can be attained for more accurate calculation. The factors listed are generic values so consult the manufacturer for specific values.

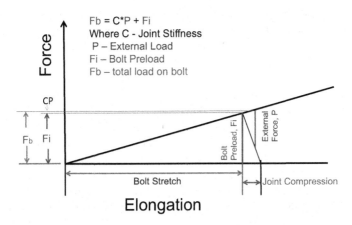

$Fb = C*P + Fi$
Where C - Joint Stiffness
  P – External Load
  Fi – Bolt Preload
  Fb – total load on bolt

FIGURE 7.1   – Joint diagram

There are other factors that can reduce bolt preload while tightening, such as a larger bolt head or oversized bolt hole, cupped washers, lack of lubrication corrosion, thread condition (stretched, cut, damaged), holes misaligned, thread contamination, and many others.

Lubricated torque is typically more accurate and more repeatable after many uses than dry torque. Some studies show as much as a 25% better bolt preload using lubricated conditions. Lubricated torque conditions also lower torque at assembly making it easier to control. Lubrication not only goes on the threads but also under the bolt head. Many will neglect the head friction during assembly which can lead to a potential 50% error in bolt preload.

A typical bolt target torque is 75% proof load on reusable joints (those that have maintenance). Permanent joints (like in structural assemblies) may be preloaded to a 90% proof load. Proof strength is typically 5–10% lower than yield strength for the material grade. The yield strength is 60–90% of tensile strength so there is a considerable margin for overloading a bolt before over-stressing most bolt grades.

There are many methods for executing bolt torque with various accuracies as shown below.

- By feel +/-35%
- Click torque wrench +/-25%
- Turn of nut +/-15%
- Elongation micrometer +/-3–5% and elongation ultrasonic +/-1%

For large bolts where bolt torques can be much larger than standard torque wrench capacity, a torque multiplier may be used. The turn of nut method is common for very large bolts when a torque multiplier is not available or feasible.

Bolt torque execution is very important. For bolt circle applications an even tightening around a bolt circle, such as a star pattern. Ramp up the bolt torque for each pass, such as 30%, 60% and a final 100% of the final target torque. On a rotating bolted element, rotate after each torque pass to even out the bolt preload.

There are several reasons for the loss of an initial bolt preload, some of which include vibration, embedment relaxation (bolt embeds into surface over time), and elastic interactions (tighten one, and the one next to it loosens as the joint compresses). Therefore, a star pattern, ramped torque and rotating assembly is a best practice. In some assemblies, there may be increased bolt friction from lifting a load during tightening. It is best to have parts in place when a bolt preload procedure is executed. One way to see how vibration resistant the joint is in performance is to record the removal torque (when compared to installation torque) of an assembly. While the breakaway torque is higher than the initial torque, inconsistencies can indicate assembly issues.

There are several items where maintaining the bolt preload can be enhanced by ensuring:

- The correct initial installation
- The external load doesn't exceed the design
- Nuts are not reused
- New screws are used
- Locknuts (Nordlock), thread locking (Loctite) are used
- The procedure is retightened after startup (warmup can loosen some joints)
- The bolt assemblies are tack welded or the bolt heads are tie-wired together

There are many bolt failure modes such as overload (stretched or stripped threads, tensile overload), fatigue, shear, corrosion, and wear. Small bolts are more vulnerable to stretch or strip threads. Coarse threads have more shear strength than fine threads (more thread area).

Tensile overload happens many times due to a high external load or as part of a secondary failure. If the joint is assembled correctly, the bolt should generally not fail in shear. When a bolt fails in shear then the joint has slipped. A low bolt preload/clamp force is typically the root cause of joint slip. It is always good to avoid bolt threads in a shear plane if possible.

Bolt failure typically occurs in three main bolt areas – 15% at head fillet, 20% at thread runout, 65% at first thread to engage nut. The first thread sees the highest stress so there is a high probability of failure there. The other two areas can be vulnerable to bending stresses that concentrate on bolt threads at those locations.

Fatigue is maybe the most common failure mode of bolts, with some studies showing 85% of fastener failures due to fatigue. Many fundamentals from the fatigue discussion apply, but there are also some special considerations for bolted joints. Formed/rolled threads are more fatigue resistant than machined threads. Eliminating head angularity to the mating surface is critical to preventing bolt fatigue. Head angularity induces high bending stress to the bolt heads. Some studies have shown drastic fatigue strength reduction if the head angularity is above 1°.

As with any fatigue failure, if the amplitude of cyclic stress can be reduced below the endurance limit of the bolts then infinite fatigue life can be attained. For bolted joints this can best be achieved by having the proper joint clamp load (bolt preload). Under this condition the joint is compressed sufficiently where the external load is mostly absorbed by joint compression and not by the bolt itself, as shown in Figure 7.2. Keep in mind the external load is cyclic going up and down in magnitude. Reduce the cyclic force and fatigue is neutralized. There are some key applications where this is a major factor, such as rotating bolted joints.

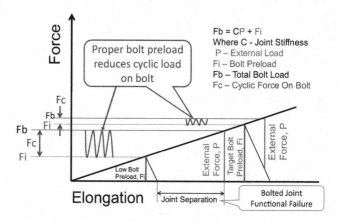

FIGURE 7.2   – Joint diagram showing cyclic load on bolts

7.2 CASE STUDY – DRYER JOURNAL BOLT FAILURES

FAILURE EVENT

A paper machine experienced five dryer journal bolt functional failures in 8 years (dryers 5, 20, 43, 47, 45). Some of the failures caused as much as 2 days' downtime at a cost of $350,000. On the last failure (dryer 45), the journal and bolts had only been installed less than a year from a previous repair. Not only are the failures creating a significant financial impact but there is also a potential safety risk as this Level 4 failure occurs.

FAILURE MODES

There were multiple failures which exhibited similar characteristics but to simplify the analysis the last failure is summarized below from dryer 45. Since this failure is a fastened joint along a rotating bolt circle, the failure mode is presented for each bolt around the bolt circle. Each bolt typically had an element of fatigue and overload. Figure 7.3 shows which failure mode was dominant for each bolt. The bolt number and neutral axis are arbitrary but are drawn to show how one side has more fatigue and the other side more overload. The midpoint of the bolt circle is also important as the joint cycles through tension and compression each 180° creating a cyclic load. All bolts were new at the last installation except #15 which was reused.

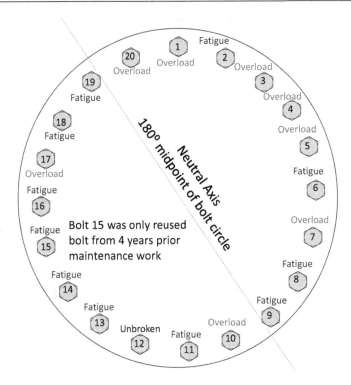

FIGURE 7.3   – Detail of bolt circle failure mode

A high-level description of the fracture surfaces is listed below.

- Primary failure – Reversed bending fatigue of screws at first engaged threads of joint which was around thread #7. Seven of the fatigue fractures were older fatigue age.
- Secondary failure – internal siphon assembly was damaged
- Secondary failure – damaged journals in fit area into dryer head and some minor thread wear was observed.

The failure mode showed no evidence of a corrosion element and no steam leaks were evident before any journal failures.

POSSIBLE CAUSES

- Torsional overload of screws on installation
- Excessive external load on dryer journal joint – felt tension, wraps-doctor loads, flooded dryers, etc.

- Insufficient bolt preload (joint clamp load)
- Non-uniform bolt preload around journal bolt circle.

ANALYSIS

The original dryer journal design had an interference fit from journal to dryer head with journal bolts attaching journal to dryer head. The manufacturer had specified 24mm grade 8.8 screws at a 157KN (32,295lbf) bolt preload. To achieve the joint clamp load, a bolt torque of 753 N-m (556 ft-lbs dry or 417 ft-lbs lubed) or.177mm (.007") of elongation was needed.

Looking at the method of differences some general observations can be made. It is somewhat obvious that there is a set of conditions that created the failures. All dryers are of the same design, yet most dryers have experienced no failures. There have been failures on top dryers and bottom dryers, but no correlation between driven dryers and those with maintenance bearing changes to the failures. All these dryer journal failures were believed to have had maintenance work previously executed.

Since the failures started occurring, the machine speeds had increased, and the steam/condensate system was changed to the drive side. Bottom dryers 2–16 and 44 and 46 were converted to vacuum dryers. The second section changed to unirun dryers where bottom dryers have the steam load taken off.

The bottom dryers have less dryer load as the felt tension is lifting on the dryers. There was only one failure on a bottom dryer, so most were top dryers, which have additional felt loading.

Dryer 45 had several instances where the sheet broke and wrapped dryer getting behind doctor which put additional bending stress on dryer. This is a potential increase in the external load on the dryer journal joint. However, all dryers that experienced failures did not have dryer doctors. It is also true that no two dryers have the same bolt preload conditions.

Flooding events showed no increase in drive torque so that either the flooded dryers were limited to only a few or the flooding was the result of a secondary failure. There are a couple of reasons why the dryer failure can be as a result of a secondary failure effect rather than a contributing cause. One is when the steam system seal is broken. When the dryer journal bolts fail then the journal causes steam joint misalignment. As the steam joint leaks, the differential pressure to evacuate the dryer is compromised and the dryer will begin to flood as condensing steam in the dryer can't escape.

The second way the dryer can be found flooded as a secondary failure is that when dryer misalignment occurs it wrecks the syphon siphon in the dryer. This removes the means by which to drain condensate from the dryer. These secondary failures (after the primary failure of journal bolts) would explain why drive torque never increased prior to bolted journal failures. Dryer 45 had both failure modes when found with the journal failure.

Either way the machine manufacturer has said that the dryer can operate under flooded conditions without the failure of the journal bolted joint. Likely this has occurred many times in various places on the machine over the years, also without failure, as flooding is common and random. At worst, the flooded dryers are a contributing cause to failures but not the root cause.

For a rotating bolted joint on a roll, the physics of failure of the joint is critical to find the root cause. As the roll rotates the bolt/joint undergoes tension and compression as bending stress reacts around the journal. Tension is at the bottom of the rotation (6 o'clock) and compression at the top (12 o'clock). So, the roll section is broken up into two halves of stress – tension and compression. The maximum stress occurs perpendicular to the neutral axis and decreases as bolts are at the neutral axis, as shown in Figure 7.4.

Looking back at Figure 7.3, which shows the dominant failure mode around the bolt circle, one half of the bolt circle shows fatigue dominance versus overload dominance. The fatigue side would likely occur first as a lower cyclic stress occurs over time. The random overloaded bolts in the fatigue region would overload once the fatigued bolts fractured and neighboring bolts took the load of many bolts in sequence through the rotating stress. The one unbroken bolt was likely very loose in operation. It could have had poor threads, become loose during failure, or just didn't get preloaded well on installation. The side with more overload failure occurred last as each successive bolt rotated through maximum tension and the bending stress carried all the incoming load as the neighboring bolts had already failed. The ultimate failure of the rotating bolt circle is like a zipper once it starts. The only way to prevent it is to never let it start to fail in one area.

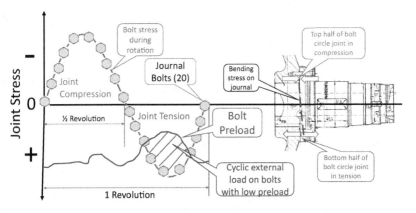

FIGURE 7.4 – Dryer journal operational stress

It is worth noting that bolt 15 was the only reused bolt from the previous failure, likely due to it not being broken or a supply issue. It could have been just one weak point that contributed to the catastrophic failure since no crack testing was done. Bolt 15 failed due to fatigue.

The rotating bolted joint operating stress is shown in Fig. 7.4 as the roll rotates from tension thru compression each revolution.

The bolts that failed ran for less than a year which at 3000fpm (191rpm) on a 60" diameter dryer would be 10^7 stress cycles, which is in the finite cycles range. The increase in speed from the machine would increase the stress cycles on the dryer joint but would not have a large impact on the cyclic stress

There was no consistent use of torque wrenches on journal bolt installations so a variable clamp load on joints was expected. After the emergency repair outage from this failure, a replacement job was scheduled to replace some journals and install new bolts. A new procedure utilizing bolt elongation and ultrasonic measurements was also utilized to ensure proper bolt preload. This was also used to measure the as-found condition from previous bolt assembly, which was critical in confirming the original failure hypothesis of a poor bolt preload.

The results of the as-found and as-installed bolt preload are shown in Figure 7.5. One half of the bolt circle had bolts with ~50% of the preload as other bolts. This would result in a significant increase in the cyclic stress to those low preload bolts, as discussed in the bolt failure key elements in Figure 7.2, and would cause a large increase in the cyclic stress on that half of the assembly. With the increased cyclic stress being more than the endurance limit of the application, fatigue failure was guaranteed.

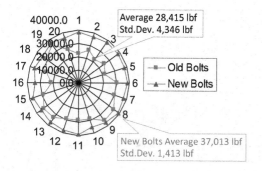

FIGURE 7.5 – Bolt preload distribution

ROOT CAUSE

The physical root cause was insufficient bolt preload and bolt preload scatter around the bolt circle on installation leading to reversed bending fatigue on the dryer journal bolts as the high cyclic stress exceeded the endurance limit for the application. The latent root cause was the lack of installation procedures for dryer journal bolts and the lack of precision installation practices.

POSSIBLE ACTION ITEMS

- Implement fastener installation best practices such as proper lubrication (threads and under head), a specified tightening pattern, no impact wrenches, use of torque wrenches and ramped up torque
- Implement an ultrasonic elongation bolt installation procedure to ensure uniform bolt preload on future high-risk journal bolt replacements
- Implement a good target bolt torque procedure for new design and proven bolt elongation methods for execution during normal maintenance times
- Redesign the journal to include increased head fit and fit length for joint insurance
- Check torque wrench calibration before dryer journal bolt installations
- Increase bolt preload and bolt material (8.8 to B7) to improve overall joint clamp load. B7 material at 400°F (204°C) has a yield strength ~92,000 psi (634MPa) vs 76,000psi (524MPa) for grade 8.8 original
- Training for mill technicians on precision bolt assembly best practices.

SUMMARY

All the high-risk dryers that had previous failures or suspected issues had journals or journal bolts replaced in the following 3 years using precision installation methods. Almost all of the bolted joints showed similar bolt preload scatter issues and from previous poor installations. No more journal failures occurred after this work was completed. The action item to investigate a dryer journal redesign to improve the service factor was investigated and designed by the manufacturer but was not pursued once the other action items had been completed with positive results.

Index

Printed in the United States
by Baker & Taylor Publisher Services